Bioenergy

Bioenergy

Josh Olson

Larsen & Keller
www.larsen-keller.com

Bioenergy
Josh Olson
ISBN: 978-1-64172-413-5 (Hardback)

 Larsen & Keller

Published by Larsen and Keller Education,
5 Penn Plaza,
19th Floor,
New York, NY 10001, USA

Cataloging-in-Publication Data

Bioenergy / Josh Olson.
    p. cm.
Includes bibliographical references and index.
ISBN 978-1-64172-413-5
1. Biomass energy. 2. Energy conversion. 3. Energy crops. 4. Refuse as fuel.
5. Waste products as fuel. I. Olson, Josh.
TP339 .B56 2020
662.88--dc23

For more information regarding Larsen and Keller Education and its products, please visit the publisher's website www.larsen-keller.com

TABLE OF CONTENTS

This book has been written, keeping in view that students want more practical information. Thus, my aim has been to make it as comprehensive as possible for the readers. I would like to extend my thanks to my family and co-workers for their knowledge, support and encouragement all along.

The renewable energy that is made from the materials that are derived from biological sources is known as bioenergy. These are organic materials that have stored sunlight in the form of chemical energy. A few of such materials are wood, straw, wood waste, manure, sugarcane and other crop residues that can be used as fuels. Various agricultural products such as corn, soybeans, willow, switchgrass and sugarcane are specifically grown for the production of biofuel. Dry and combustible agricultural products such as wood pellets can be burned to boil water, and the steam generated is used to drive a turbine to produce electricity. Municipal and household waste can also be used as a source of bioenergy. This book elucidates new techniques and their applications in a multidisciplinary approach. This book explores all the important aspects of bioenergy in the present day scenario. It will provide comprehensive knowledge to the readers.

A brief description of the chapters is provided below for further understanding:

Chapter – What is Bioenergy?

The renewable energy that is derived from the materials obtained from biological sources is known as bioenergy. A few of its sources include wood waste, straw, wood and other crop residues. This is an introductory chapter which will introduce briefly all the significant aspects of bioenergy as well as its pros and cons.

Chapter – Basic Concepts of Bioenergy

Some of the basic concepts of bioenergy are energy crops, energy forestry, energy content of biofuel, energy returned on energy invested and cellulosic ethanol commercialization. This chapter discusses in detail these key concepts related to bioenergy.

Chapter – Sources of Bioenergy

Some of the most common sources of bioenergy are biomass, natural gas, biogas, biochar, bioliquids, renewable natural gas, bagasse, wood gas, biofuel and wood fuel. The chapter closely examines these fundamental sources of bioenergy to provide an extensive understanding of the subject.

Chapter – Technologies for Bioenergy Production

The important techniques used to derive bioenergy from different sources include biorefinery, bioconversion, thermal depolymerization, Fischer–Tropsch Process, biomass heating system, and carbon capture and storage. All these diverse bioenergy techniques have been carefully analyzed in this chapter.

Chapter – Bioenergy Feedstocks

The biological materials which can be used directly as a fuel or converted to another form of fuel or energy product are called feedstocks. They can be classified as starch-based feedstocks, oilseed-based feedstocks, fiber and grass cellulosic feedstocks and algae-based feedstocks. This chapter has been carefully written to provide an easy understanding of these types of bioenergy feedstocks.

Chapter – Biofuels

The fuel which is produced using modern techniques and processes from biomass is referred to as biofuel. Some of the various types of biofuels are algal fuel, methanol fuel, ethanol fuel, butanol fuel, pellet fuel, biodiesel, etc. All these diverse types of biofuels have been carefully analyzed in this chapter.

Josh Olson

What is Bioenergy?

CHAPTER 1

The renewable energy that is derived from the materials obtained from biological sources is known as bioenergy. A few of its sources include wood waste, straw, wood and other crop residues. This is an introductory chapter which will introduce briefly all the significant aspects of bioenergy as well as its pros and cons.

RENEWABLE ENERGY

Renewable energy uses energy sources that are continually replenished by nature—the sun, the wind, water, the Earth's heat, and plants. Renewable energy technologies turn these fuels into usable forms of energy—most often elec- tricity, but also heat, chemicals, or mechanical power.

Importance of Renewable Energy

Today we primarily use fossil fuels to heat and power our homes and fuel our cars. It's convenient to use coal, oil, and natural gas for meeting our energy needs, but we have a limited supply of these fuels on the Earth. We're using them much more rapidly than they are being created. Eventually, they will run out. And because of safety concerns and waste disposal prob- lems, the United States will retire much of its nuclear capacity by 2020. In the mean- time, the nation's energy needs are expected to grow by 33 percent during the next 20 years. Renewable energy can help fill the gap.

A PV-system at the Pinnacles National Monument in California eliminates a $20,000 annual fuel bill for a diesel generator that pro- duced each year 143 tons of carbon dioxide—a greenhouse gas.

Even if we had an unlimited supply of fos- sil fuels, using renewable energy is better for the envi- ronment. We often call renew- able energy technologies "clean" or "green" because they produce few if any pollutants. Burning fossil fuels, however, sends greenhouse gases into the atmos- phere, trapping the sun's heat and con- tributing to global warming. Climate scientists generally agree that the Earth's average temperature has risen in the past century. If this trend continues, sea levels will rise, and scientists predict that floods, heat waves, droughts, and other extreme weather conditions could occur more often.

Other pollutants are released into the air, soil, and water when fossil fuels are burned. These pollutants take a dramatic toll on the environment—and on humans. Air pollution contributes to diseases like asthma. Acid rain from sulfur dioxide and nitrogen oxides harms plants and fish. Nitrogen oxides also contribute to smog.

Renewable energy will also help us develop energy independence and security. The United States imports more than 50 percent of its oil, up from 34 percent in 1973. Replacing some of our petroleum with fuels made from plant matter, for example, could save money and strengthen our energy security.

Renewable energy is plentiful, and the technologies are improving all the time. There are many ways to use renewable energy. Most of us already use renewable energy in our daily lives.

BIOENERGY

Bioenergy is renewable energy created from natural, biological sources. Many natural sources, such as plants, animals, and their byproducts, can be valuable resources. Modern technology even makes landfills or waste zones potential bioenergy resources. It can be used to be a sustainable power source, providing heat, gas, and fuel.

Because the energy contained in sources like plants is obtained from the sun through photosynthesis, it can be replenished and is considered an inexhaustible source.

Using bioenergy has the potential to decrease our carbon footprint and improve the environment. While bioenergy uses the same amount of carbon dioxide as traditional fossil fuels, the impact can be minimized as long as the plants used are replaced. Fast-growing trees and grass help this process and are known as bioenergy feedstocks.

Most bioenergy comes from forests, agricultural farms, and waste. The feedstocks are grown by farms specifically for their use as an energy source. Common crops include starch or sugar-based plants, like sugarcane or corn.

To turn raw sources into energy, there are three processes: chemical, thermal, and biochemical. Chemical processing uses chemical agents to break down the natural source and convert it into liquid fuel. Corn ethanol, a fuel created from corn, is an example of chemical processing results. Thermal conversion uses heat to change the source into energy through combustion or gasification. Biochemical conversion uses bacteria or other organisms to convert the source, such as through composting or fermentation.

Bioenergy exists at several different levels. Individuals can create bioenergy, such as by creating a compost heap out of kitchen scraps and keeping worms to produce rich fertilizer. At the other extreme are large energy corporations looking for more sustainable energy sources than oil or coal. These organizations use huge farms and facilities to provides hundreds or thousands of customers with energy.

Importance of Bioenergy

Having the ability to produce energy through plants or other resources can lessen U.S. reliance on foreign nations for sources of energy. Bioenergy also is viewed as essential for the environment. Continued use of fossil fuels can cause significant environmental issues by producing greenhouse gases that contribute to global warming or by emitting harmful pollutants such as sulfur dioxide which can harm the population's health.

As technology progresses, bioenergy has the potential to dramatically reduce greenhouse emissions, the release of harmful gases associated with global warming and climate change. The use of forests and farms in bioenergy can help combat the harmful release of carbon dioxide and help achieve a balance.

At this time, bioenergy is not ready to replace fossil fuels. The process is too costly and uses too many resources to be practical in most areas. The large plots of land and significant amounts of water needed to be successful can be difficult for many states or countries. Additionally, agricultural resources such as land and water dedicated to producing crops related to bioenergy can limit the resources used for producing food. Still, as science continues to study this area, bioenergy could increasingly become a larger source of energy that can help improve the environment.

Solid Biomass

Simple use of biomass fuel
(Combustion of wood for heat).

Sugarcane plantation to produce ethanol.

One of the advantages of biomass fuel is that it is often a by-product, residue or waste-product of other processes, such as farming, animal husbandry and forestry. In theory this means there is no competition between fuel and food production, although this is not always the case. Land use, existing biomass industries and relevant conversion technologies must be considered when evaluating suitability of developing biomass as feedstock for energy.

Biomass is the material derived from recently living organisms, which includes plants, animals and their byproducts. Manure, garden waste and crop residues are all sources of biomass. It is a renewable energy source based on the carbon cycle, unlike other natural resources such as petroleum, coal, and nuclear fuels. Another source includes Animal waste, which is a persistent and unavoidable pollutant produced primarily by the animals housed in industrial-sized farms.

A CHP power station using wood to supply 30,000 households.

There are also agricultural products specifically being grown for biofuel production. These include corn, and soybeans and to some extent willow and switchgrass on a pre-commercial research level, primarily in the United States; rapeseed, wheat, sugar beet, and willow (15,000 ha or 37,000 acres in Sweden) primarily in Europe; sugarcane in Brazil; palm oil and miscanthus in Southeast Asia; sorghum and cassava in China; and jatropha in India. Hemp has also been proven to work as a biofuel. Biodegradable outputs from industry, agriculture, forestry and households can be used for biofuel production, using e.g. anaerobic digestion to produce biogas, gasification to produce syngas or by direct combustion. Examples of biodegradable wastes include straw, timber, manure, rice husks, sewage, and food waste. The use of biomass fuels can therefore contribute to waste management as well as fuel security and help to prevent or slow down climate change, although alone they are not a comprehensive solution to these problems.

Biomass can be converted to other usable forms of energy like methane gas or transportation fuels like ethanol and biodiesel. Rotting garbage, and agricultural and human waste, all release methane gas—also called "landfill gas" or "biogas." Crops, such as corn and sugar cane, can be fermented to produce the transportation fuel, ethanol. Biodiesel, another transportation fuel, can be produced from left-over food products like vegetable oils and animal fats. Also, Biomass to liquids (BTLs) and cellulosic ethanol are still under research.

Sewage Biomass

The use of municipal and household waste is on the forefront of new sources for biomass, and is a largely discarded resource on which new research is being conducted for use of energy production. A new bioenergy sewage treatment process aimed at developing countries is now on the horizon; the Omni Processor is a self-sustaining process which uses the sewerage solids as fuel to convert sewage waste water into drinking water and electrical energy. Sewage sludge is a point of focus in current research for developing bioenergy from biomass. The large quantity being produced by households at a continuous rate presents an opportunity to extract valuable compounds contained within it which can be then used to produce bioenergy. The main form of bioenergy being produced from sewage is methane, but producing other forms is still being researched. The use of sewage to produce methane reduces the amount of waste put into landfills, its costs of transportation and disposal, and also keeps a larger amount of gas out of the atmosphere, as more is able to be captured.

Electricity Generation from Biomass

The biomass used for electricity production ranges by region. Forest byproducts, such as wood residues, are popular in the United States. Agricultural waste is common in Mauritius (sugar cane residue) and Southeast Asia (rice husks). Animal husbandry residues, such as poultry litter, is popular in the UK.

Electricity from Sugarcane Bagasse

Sugarcane (*Saccharum officinarum*) plantation ready for harvest.

Sucrose accounts for little more than 30% of the chemical energy stored in the mature plant; 35% is in the leaves and stem tips, which are left in the fields during harvest, and 35% are in the fibrous material (bagasse) left over from pressing.

The production process of sugar and ethanol in takes full advantage of the energy stored in sugarcane. Part of the bagasse is currently burned at the mill to provide heat for distillation and electricity to run the machinery. This allows ethanol plants to be energetically self-sufficient and even sell surplus electricity to utilities; current production is 600 MW (800,000 hp) for self-use and 100 MW (130,000 hp) for sale. This secondary activity is expected to boom now that utilities have been induced to pay "fair price "(about US$10/GJ or US$0.036/kWh) for 10 year contracts. This is approximately half of what the World Bank considers the reference price for investing in similar projects. The energy is especially valuable to utilities because it is produced mainly in the dry season when hydroelectric dams are running low. Estimates of potential power generation from bagasse range from 1,000 to 9,000 MW (1,300,000 to 12,100,000 hp), depending on technology. Higher estimates assume gasification of biomass, replacement of current low-pressure steam boilers and turbines by high-pressure ones, and use of harvest trash currently left behind in the fields. For comparison, Brazil's Angra I nuclear plant generates 657 MW (881,000 hp).

Presently, it is economically viable to extract about 288 MJ of electricity from the residues of one tonne of sugarcane, of which about 180 MJ are used in the plant itself. Thus a medium-size distillery processing 1,000,000 tonnes (980,000 long tons; 1,100,000 short tons) of sugarcane per year could sell about 5 MW (6,700 hp) of surplus electricity. At current prices, it would earn US$18 million from sugar and ethanol sales, and about US$1 million from surplus electricity sales. With advanced boiler and turbine technology, the electricity yield could be increased to 648 MJ per tonne of sugarcane, but current electricity prices do not justify the necessary investment. According to one report, the World Bank would only finance investments in bagasse power generation if the price were at least US$19/GJ or US$0.068/kWh.

A sugar/ethanol plant located in Piracicaba, São Paulo State. This plant produces the electricity it needs from bagasse residuals from sugarcane left over by the milling process, and it sells the surplus electricity to the public grid.

Bagasse burning is environmentally friendly compared to other fuels like oil and coal. Its ash content is only 2.5% (against 30–50% of coal), and it contains very little sulfur. Since it burns at relatively low temperatures, it produces little nitrous oxides. Moreover, bagasse is being sold for use as a fuel (replacing heavy fuel oil) in various industries, including citrus juice concentrate, vegetable oil, ceramics, and Tyre Recycling. The state of São Paulo alone used 2,000,000 tonnes (1,970,000 long tons; 2,200,000 short tons), saving about US$35 million in fuel oil imports.

Researchers working with cellulosic ethanol are trying to make the extraction of ethanol from sugarcane bagasse and other plants viable on an industrial scale.

Electricity from Electrogenic Micro-organisms

Another form of bioenergy can be attained from microbial fuel cells, in which chemical energy stored in wastewater or soil is converted directly into electrical energy via the metabolic processes of electrogenic micro-organisms. The power generation capability of this technology has not been found to be economically viable till date, however, this technology has found been found to be more useful for chemical treatment processes and student education.

Environmental Impact

Some forms of forest bioenergy have recently come under fire from a number of environmental organizations, including Greenpeace and the Natural Resources Defense Council, for the harmful impacts they can have on forests and the climate. Greenpeace recently released a report entitled Fuelling a BioMess which outlines their concerns around forest bioenergy. Because any part of the tree can be burned, the harvesting of trees for energy production encourages Whole-Tree Harvesting, which removes more nutrients and soil cover than regular harvesting, and can be harmful to the long-term health of the forest. In some jurisdictions, forest biomass is increasingly consisting of elements essential to functioning forest ecosystems, including standing trees, naturally disturbed forests and remains of traditional logging operations that were previously left in the forest. Environmental groups also cite recent scientific research which has found that it can take many decades for the carbon released by burning biomass to be recaptured by regrowing trees, and even longer in low productivity areas; furthermore, logging operations may disturb forest soils and cause them to release stored carbon. In light of the pressing need to reduce greenhouse gas emissions in the short term in order to mitigate the effects of climate change, a number of environmental groups are opposing the large-scale use of forest biomass in energy production.

Suppose you cut down a 50-year oak tree in your garden and use the logs to heat your house instead of coal. Wood emits more carbon dioxide than coal per unit of heat gained and the roots left in the soil emit more carbon dioxide as they rot. If you plant another tree it will soak up that carbon dioxide in about 50 years. But if you had left the original tree in place it would have soaked up the carbon dioxide from the coal and more. It could take centuries before cutting down the tree would give any benefit. But the world needed to cut carbon dioxide over the next few decades if the global warming was to be kept below 3 degrees C. The journal also concluded that official claimed carbon reductions from renewables had been overstated. The European Union, for example, got more 64% of its renewable energy from biomass (mostly wood) but United Nations and EU rules did not count the carbon emissions from burning biomass.

Recently, a new company called Mango materials used bacterial fermentation to produce an intracellular biopolymer, polyhydroxyalkanoate from methane. The great advantage of biopolymers is that it is biodegradable which makes it environment friendly. Because methane is being used that decreases the price of polymers that it would compete with traditional plastics. Also, because methane would be converted into biopolymer that would reduce methane emissions. Chief Executive Officer Molly Morse said that the unused methane would be enough to produce more than three billion pounds of biopolymer. Morse announced in 2017 that using this polymer will reduce the waste in the textile industry because it will be reproduced as biopolymer again in every effective industrial loop.

Pros and Cons of Bioenergy

Pros

- Bioenergy a reliable source of renewable energy. We will never have a shortage of waste that can be converted to energy. As long as there is garbage, manure, and crops there will be biomass to create bioenergy.

- Bioenergy can be stored with little energy loss.

- As long as there is agriculture there will be a constant energy source.

- Bioenergy emits little or no greenhouse gas emissions and is carbon neutral. The carbon that is created by biomass is reabsorbed by the next crop of plants.

- Bioenergy doubles as a waste disposal measure.

- Bioenergy crops help stabilize soils, improve soil fertility, and reduces erosion.

- Bioenergy is a source of clean energy, the use of which can result in tax credits from the US government.

- Bioenergy reduces the need for landfills.

Cons

- Using wood from natural forests can lead to deforestation if the forests are not replanted.

- The cost of harvesting, transporting, and handling biomass can be expensive.

- Storing and processing of biomass requires large amounts of space.

- Some fuel sources are seasonal.

- May compete with food production in specific cases.

References

- What-is-bioenergy-2941107: thebalance.com, Retrieved 20 July, 2019

- Cho, Renee (2011-08-18). "Is Biomass Really Renewable?". Earth Institute. Columbia University. Retrieved 2016-10-01

- Le Page, Michael (2016-09-24). "The Great Carbon Scam". The New Scientist. 3092. 231 (3092): 20–21. doi:10.1016/S0262-4079(16)31736-5

- "Janicki Bioenergy website". Archived from the original on 9 January 2015. Retrieved 11 January 2015

- Pros-and-cons-of-bioenergy: alcse.org, Retrieved 25 August, 2019

- "NRDC fact sheet lays out biomass basics, campaign calls for action to tell EPA our forests aren't fuel". nrdc.org. Archived from the original on 4 October 2013. Retrieved 28 February 2015

Basic Concepts of Bioenergy

Some of the basic concepts of bioenergy are energy crops, energy forestry, energy content of bio-fuel, energy returned on energy invested and cellulosic ethanol commercialization. This chapter discusses in detail these key concepts related to bioenergy.

ENERGY CROP

Human beings run on energy and to run our bodies, we use food as fuel. It may seem ironic, then, that in a bid to harness more kinds of natural and renewable energy, certain plants are grown with the single-minded agenda of turning them into biofuel or combusting them to generate energy. Such plants are dubbed 'energy crops.'

Short Rotation Energy Crops

There are two kinds of short rotation energy crops, depending on the cycle and scale. In short rotation coppice (SRC), fast-growing young trees are cut at the stump every winter when they lie dormant, thus giving rise to many new stems in the growing season (the yield stands at 2-4 years). Trees like poplar and willow are popular choices for SRC. Short Rotation forestry is planting a site and then felling the tree when their stems reach a diameter of 10-20cm, at chest-height. This cycle is longer and takes place every 8-20 years; popular plants include sycamore, ash, poplar, eucalyptus and beech.

Non-woody Crops and Grasses

Miscanthus is the most popular of all non-woody energy crops and gives a good annual yield as opposed to short rotation energy crops that give yield only every 2-20 years. Other potential energy crops include hemp and varieties of reed, rye and switchgrass (from the grasses family) too can be considered. With these plants though, there's always the threat of invasion so these need to be taken up with care and caution.

Agricultural Energy Crops

By virtue of being high in carbon content, many conventional crops such as sugar crops (sugar beet), starch crops (wheat, maize, and potatoes) and oil crops (rapeseed, waste vegetable oil) are either used straight as fuel or hydrolyzed into biofuel.

Aquatic Crops (Hydroponics)

The advantage of aquatic crops is that they don't use land and they take all that they need in form of nutrients from the water, thus being excellent in photosynthesis. Algae, both microscopic and macroscopic (such as seaweed, kelp) and other pond and lake flora are good forms of energy crops – the disadvantage lies in their high water content which needs to be dried up before being used as biomass.

Grain and Seed Crops

The presence of wheat and barley seed in archaeological sites dating back to 6750 BC is testimony to the importance that cereals have played in the devel- opment of human societies. Today, maize, rice and wheat dominate world agricultural production and, together with a whole range of other grain and seed crops, provide the staple food of populations worldwide.

Grain and seed crops have also traditionally been used in fermentation to produce beer, wine and spirits, because the stored carbohydrates (sucrose and starch) can be readily broken down by enzymatic systems. The adoption of this process to produce bioethanol for vehicles, however, has resulted in the use of food grain for nonfood purposes on an unprecedented scale. Maize and wheat currently make the largest contribution to biofuel production, matched only by that from sugarcane and oil palm.8 Maize is also used for biogas. In addition, the parts of cereal crops that are not used for food (e.g. wheat straw and corn stover) can be used as a source of biomass for thermal conversion and lignocellulose for second- generation biological conversion. Similarly, grain from sorghum as well as stalks of sweet sorghum can both be used for biofuels, whilst sorghum stovers could be used as a source of lignocellulose.

Sugar Crops

Sugars are transported in plant stems in normal development but some species can also store high concentrations. The main sugar crops used for bioenergy are sugarcane, sugar beet and, , sweet sorghum.

Bioethanol production from sugarcane is an extremely efficient and well- developed industry in Brazil. Indeed, it is interesting to contemplate whether or not other biofuel crops would really be competitive if sugarcane could be cultivated throughout the world in the way that it can in the tropics and subtropics. Most certainly, it is among the most productive plants known and it is also able to store high concentrations of sucrose in the stem. In addition, sugarcane bagasse (the fibrous residue) is a primary fuel source, making most sugarcane mills extremely efficient. It could also be a source of lignocelluloses and the feasibility of producing ethanol from bagasse is cur- rently under investigation. Sweet sorghum is adapted to both humid and tro- pical climates but can be grown in colder climates than sugarcane. In cooler temperate climates, sugar beet can be used as a source of sugar for bioethanol.

Oil Crops

Oil palm is by far the largest producer of oil for biodiesel. However, a large range of alternative oil crops are grown in areas where the climate does not favour oil-palm production. This includes soybean (parti- cularly in the Americas) and oilseed rape (particularly in Europe and cooler temperate areas). More recently, Jatropha has been heralded as a promising biofuel crop for drought-prone environments, as well as other species, such as Pongamia. Other oil crops, such as sunflower, babassu palm, peanut and even olive are also used for biodiesel.

Dedicated Biomass Crops

As pointed out previously, biomass can be obtained from any crop. Indeed, practices such as using the whole wheat crop (grain included) for combustion are known to have been carried out (this probably exacerbates global warming through the production of nitrogen oxides, which are much more potent GHGs than CO_2). However, the term "dedicated biomass crops" refers to nonfood crops that are solely grown for biomass production. These comprise mostly perennial grasses and fast-growing trees. Dedicated biomass crops were first developed for combustion and thermal conversion technologies but, due to their potential to supply high yields of lignocelluloses, have become of interest for second-generation biofuels. Perennial grasses are also widely used for biogas, but wood chip is not suitable for this process.

An impressive number of perennial grasses are used as energy crops but, in this volume, coverage is restricted to the major ones; Miscanthus,; switchgrass and reeds. Similarly, this volume covers two main fast growing trees: willows and poplars.

Algae

Algae fall into two main types: microalgae (phytoplankton, microphytes or planktonic algae) and macroalgae (seaweed). Both are used for biofuel pro- duction, although microalgae have received most attention due to their ability to be grown in ponds and bioreactors. Macroalgae can be grown on ropes. As the photosynthetic efficiency of algae (6–8% on average) is higher than ter- restrial plants (1.8–2.2% on average) they are able to accumulate biomass at faster rates. Other advantages are that they do not require the use of high-grade productive land and can utilise a wide range of water sources (fresh, brackish, saline and waste water).

Advantages of Energy Crops

Growing agriculture for energy means lesser dependence on fossil fuels and thus a decrease in mining and exports. Another big advantage is that biomass is freely available. Growing plants or a plantation means that they use up carbon dioxide and release oxygen, thus leading to a more oxygenated atmosphere. In developed countries where there are huge landfills built to handle agricultural waste, biomass energy could come in handy, thus freeing that land for other and more beneficial purposes. The government also provides many grants for farmers interested in the same.

Disadvantages of Energy Crops

The land used for the agriculture of biomass may in fact be needed to grow consumable crops – or even as land for housing or for recreational or commercial purposes. Also, to be able to sustain energy crops on a large scale, the costs associated with the conversion of the biomass into fuel need to be brought down.

Lastly the one big reason that makes energy crops stand at a disadvantage is that sometimes the conversion causes environmental pollution.

ENERGY FORESTRY

Energy forestry is a form of forestry in which a fast-growing species of tree or woody shrub is grown specifically to provide biomass or biofuel for heating or power generation.

The two forms of energy forestry are short rotation coppice and short rotation forestry:

- Short rotation coppice may include tree crops of poplar, willow or eucalyptus, grown for two to five years before harvest.

- Short rotation forestry are crops of alder, ash, birch, eucalyptus, poplar, and sycamore, grown for eight to 20 years before harvest.

Benefits

- The main advantage of using "grown fuels", as opposed to fossil fuels such as coal, natural gas and oil, is that while they are growing they absorb the near-equivalent in carbon dioxide (an important greenhouse gas) to that which is later released in their burning. In comparison, burning fossil fuels increases atmospheric carbon unsustainably, by using carbon that was added to the Earth's carbon sink millions of years ago. This is a prime contributor to climate change.

- According to the FAO, compared to other energy crops, wood is among the most efficient sources of bioenergy in terms of quantity of energy released by unit of carbon emitted. Other advantages of generating energy from trees, as opposed to agricultural crops, are that trees do not have to be harvested each year, the harvest can be delayed when market prices are down, and the products can fulfil a variety of end-uses.

- Yields of some varieties can be as high as 12 oven dry tonnes every year. However, commercial experience on plantations in Scandinavia have shown lower yield rates.

- These crops can also be used in bank stabilisation and phytoremediation. In fact, experiments in Sweden with willow plantations have proved to have many beneficial effects on the soil and water quality when compared to conventional agricultural crops (such as cereal).

Problems

- Although in many areas of the world government funding is still required to support large scale development of energy forestry as an industry, it is seen as a valuable component of the renewable energy network and will be increasingly important in the future.

- Growing trees is relatively water intensive.

- The system of energy forestry has faced criticism over food vs. fuel, whereby it has become financially profitable to replace food crops with energy crops. It has to be noted, however, that such energy forests do not necessarily compete with food crops for highly productive land as they can be grown on slopes, marginal, or degraded land as well - sometimes even with long-term restoration purposes in mind.

ENERGY RETURNED ON ENERGY INVESTED

In energy economics and ecological energetics, energy returned on energy invested (EROEI or ERoEI), or energy return on investment (EROI), is the ratio of the amount of usable energy (the *exergy*) delivered from a particular energy resource to the amount of exergy used to obtain that energy resource.

Arithmetically the EROEI can be defined as:

$$EROEI = \frac{\text{Energy Delivered}}{\text{Energy Required to Deliver that Energy}}$$

When the EROEI of a source of energy is less than or equal to one, that energy source becomes a net "energy sink", and can no longer be used as a source of energy, but depending on the system might be useful for energy storage (for example a battery). A related measure Energy Stored On Energy Invested (ESOEI) is used to analyse storage systems.

To be considered viable as a prominent fuel or energy source a fuel or energy must have an EROEI ratio of at least 3:1.

Application to Various Technologies

Photovoltaic

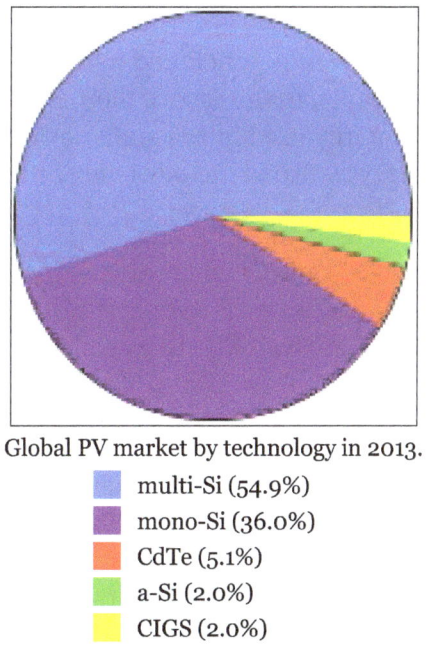

Global PV market by technology in 2013.

■ multi-Si (54.9%)
■ mono-Si (36.0%)
■ CdTe (5.1%)
■ a-Si (2.0%)
■ CIGS (2.0%)

The issue is still subject of numerous studies, and prompting academic argument. That's mainly because the "energy invested" critically depends on technology, methodology, and system boundary assumptions, resulting in a range from a maximum of 2000 kWh/m² of module area down to a minimum of 300 kWh/m² with a median value of 585 kWh/m² according to a meta-study.

Regarding output, it obviously depends on the local insolation, not just the system itself, so assumptions have to be made.

Some studies include in their analysis that photovoltaic produce electricity, while the invested energy may be lower grade primary energy.

A 2015 review in Renewable and Sustainable Energy Reviews assessed the energy payback time and EROI of a variety of PV module technologies. In this study, which uses an insolation of 1700 kWh/m²/yr and a system lifetime of 30 years, mean harmonized EROIs between 8.7 and 34.2 were found. Mean harmonized energy payback time varied from 1.0 to 4.1 years.

Wind Turbines

The EROI of wind turbines depends on invested energy in the turbine, produced energy, and life span of a turbine. In the scientific literature EROIs normally vary between 20 and 50.

Oil Sands

Because much of the energy required for producing oil from oil sands (bitumen) comes from low value fractions separated out by the upgrading process, there are two ways to calculate EROEI, the higher value given by considering only the external energy inputs and the lower by considering all energy inputs, including self generated. One study found that in 1970 oil sands net energy returns was about 1.0 but by 2010 had increased to about 5.23.

Non-manmade Energy Inputs

The natural or primary energy sources are not included in the calculation of energy invested, only the human-applied sources. For example, in the case of biofuels the solar insolation driving photosynthesis is not included, and the energy used in the stellar synthesis of fissile elements is not included for nuclear fission. The energy returned includes only human usable energy and not wastes such as waste heat.

Nevertheless, heat of any form can be counted where it is actually used for heating. However the use of waste heat in district heating and water desalination in cogeneration plants is rare, and in practice it is often excluded in EROEI analysis of energy sources.

Competing Methodology

In a 2010 paper by Murphy and Hall, the advised extended boundary protocol, for all future research on EROI, was detailed. In order to produce, what they consider, a more realistic assessment and generate greater consistency in comparisons, than what Hall and others view as the "weak points" in a competing methodology. In more recent years however a source of continued controversy is the creation of a different methodology endorsed by certain members of the IEA which for example most notably in the case of photovoltaic solar panels, controversially generates more favorable values.

In the case of photovoltaic solar panels, the IEA method tends to focus on the energy used in the factory process alone. In 2016, Hall observed that much of the published work in this field is produced by advocates or persons with a connection to business interests among the competing technologies, and that government agencies had not yet provided adequate funding for rigorous analysis by more neutral observers.

Relationship to Net Energy Gain

EROEI and *Net energy (gain)* measure the same quality of an energy source or sink in numerically different ways. Net energy describes the amounts, while EROEI measures the ratio or efficiency of the process. They are related simply by:

$$GrossEnergyYield \div EnergyExpended = EROEI$$

or

$$(NetEnergy \div EnergyExpended) + 1 = EROEI$$

For example, given a process with an EROEI of 5, expending 1 unit of energy yields a net energy gain of 4 units. The break-even point happens with an EROEI of 1 or a net energy gain of 0. The time to reach this break-even point is called energy payback period (EPP) or energy payback time (EPBT).

Economic Influence

Although many qualities of an energy source matter (for example oil is energy-dense and transportable, while wind is variable), when the EROEI of the main sources of energy for an economy fall that energy becomes more difficult to obtain and its relative price may increase.

In regard to fossil fuels, when oil was originally discovered, it took on average one barrel of oil to find, extract, and process about 100 barrels of oil. The ratio, for discovery of fossil fuels in the United States, has declined steadily over the last century from about 1000:1 in 1919 to only 5:1 in the 2010s.

Since the invention of agriculture, humans have increasingly used exogenous sources of energy to multiply human muscle-power. Some historians have attributed this largely to more easily exploited (i.e. higher EROEI) energy sources, which is related to the concept of energy slaves. Thomas Homer-Dixon argues that a falling EROEI in the Later Roman Empire was one of the reasons for the collapse of the Western Empire in the fifth century CE. In "The Upside of Down" he suggests that EROEI analysis provides a basis for the analysis of the rise and fall of civilisations. Looking at the maximum extent of the Roman Empire, (60 million) and its technological base the agrarian base of Rome was about 1:12 per hectare for wheat and 1:27 for alfalfa (giving a 1:2.7 production for oxen). One can then use this to calculate the population of the Roman Empire required at its height, on the basis of about 2,500–3,000 calories per day per person. It comes out roughly equal to the area of food production at its height. But ecological damage (deforestation, soil fertility loss particularly in southern Spain, southern Italy, Sicily and especially north Africa) saw a collapse in the system beginning in the 2nd century, as EROEI began to fall. It bottomed in 1084 when Rome's population, which had peaked under Trajan at 1.5 million, was only 15,000.

Evidence also fits the cycle of Mayan and Cambodian collapse too. Joseph Tainter suggests that diminishing returns of the EROEI is a chief cause of the collapse of complex societies, which has been suggested as caused by peak wood in early societies. Falling EROEI due to depletion of high quality fossil fuel resources also poses a difficult challenge for industrial economies, and could potentially lead to declining economic output and challenge the concept (which is very recent when considered from a historical perspective) of perpetual economic growth.

Tim Garrett links EROEI and inflation directly, based on a thermodynamic analysis that links current world energy consumption (Watts) to a historical accumulation of inflation-adjusted global wealth (US dollars) known as the Garrett Relation. This economic growth model indicates that global EROEI is the inverse of global inflation over a given time interval. Because the model aggregates supply chains globally, local EROEI is outside its scope.

Criticism of EROEI

Measuring energy output is a solved problem,
measuring the input remains highly debated.

EROEI is calculated by dividing the energy output by the energy input. Measuring total energy output is often easy, especially in the case for an electrical output where some appropriate electricity meter can be used. However, researchers disagree on how to determine energy input accurately and therefore arrive at different numbers for the same source of energy.

How deep should the probing in the supply chain of the tools being used to generate energy go? For example, if steel is being used to drill for oil or construct a nuclear power plant, should the energy input of the steel be taken into account? Should the energy input into building the factory being used to construct the steel be taken into account and amortized? Should the energy input of the roads which are used to ferry the goods be taken into account? What about the energy used to cook the steelworkers' breakfasts? These are complex questions evading simple answers. A full accounting would require considerations of opportunity costs and comparing total energy expenditures in the presence and absence of this economic activity.

However, when comparing two energy sources a standard practice for the supply chain energy input can be adopted. For example, consider the steel, but don't consider the energy invested in factories deeper than the first level in the supply chain. It is in part for these fully encompassed systems reasons, that in the conclusions of Murphy and Hall's paper in 2010, a EROI of 5 by their extended methodology is considered necessary to reach the minimum threshold of sustainability, while a value of 12-13 by Hall's methodology is considered the minimum value necessary for technological progress and a society supporting high art.

Richards and Watt propose an *Energy Yield Ratio* for photovoltaic systems as an alternative to EROEI (which they refer to as *Energy Return Factor*). The difference is that it uses the design lifetime of the system, which is known in advance, rather than the actual lifetime. This also means that it can be adapted to multi-component systems where the components have different lifetimes.

Another issue with EROI that many studies attempt to tackle is that the energy returned can be in different forms, and these forms can have different utility. For example, electricity can be converted more efficiently than thermal energy into motion, due to electricity's lower entropy. In addition, the form of energy of the input can be completely different from the output. For example, energy in the form of coal could be used in the production of ethanol. This might have an EROEI of less than one, but could still be desirable due to the benefits of liquid fuels (assuming the latters are not used in the processes of extraction and transformation).

Additional EROEI Calculations

There are three prominent expanded EROEI calculations, they are point of use, extended and societal. Point of Use EROEI expands the calculation to include the cost of refining and transporting the fuel during the refining process. Since this expands the bounds of the calculation to include more production process EROEI will decrease. Extended EROEI includes point of use expansions as well as including the cost of creating the infrastructure needed for transportation of the energy or fuel once refined. Societal EROI is a sum of all the EROEIs of all the fuels used in a society or nation. A societal EROI has never been calculated and researchers believe it may currently be impossible to know all variables necessary to complete the calculation, but attempted estimates have been made for some nations. Calculations done by summing all of the EROEIs for domestically produced and imported fuels and comparing the result to the Human Development Index (HDI), a tool often used to understand well-being in a society. According to this calculation, the amount of energy a society has available to them increases the quality of life for the people living in that country, and countries with less energy available also have a harder time satisfying citizens' basic needs. This is to say that societal EROI and overall quality of life are very closely linked.

EROEI and Payback Periods of Some Types of Power Plants

The following table is comprised from a compilation of sources of different quality. The minimum requirement is a breakdown of the cumulative energy expenses according to material data. Frequently in literature harvest factors are reported, for which the origin of the values is not completely transparent. These are not included in this table.

Type	EROEI	Amortization period	Amortization period compared to an 'ideal' power station	
			EROEI	Amortization period
Nuclear power a)				
Pressurized water reactor, 100 % [enrichment]	106	2 Months	315	17 Days
Pressurized water reactor, 83 % [enrichment]	75	2 Months	220	17 Days
Fossil energy a)				

Brown coal, Open-cast	31	2 Months	90	23 Days
Black coal, underground mining without coal transportation	29	2 Months	84	19 Days
Gas (CCGT), Natural gas	28	9 Days	81	3 Days
Gas (CCGT), Bio gas	3,5	12 Days	10	3 Days
Hydropower				
River hydroelectric	50	1 Year	150	8 Months
Solar thermal b)				
Desert, parabolic troughs + phenyl compounds medium	21	1.1 Years	62	4 Months
Wind energy b)				
1,5 MW (E-66), 2000 Full load hours VLh (German coast)	16	1.2 Years	48	5 Months
(E-66), 2700 Full load hours VLh (German coast), shore)	21	0.9 Years	63	3.7 Months
2,3 MW (E-82), 3200 Full load hours VLh (German coast), shore)	51	4.7 Months	150	1.6 Months
200 MW park (5 MW installation), 4400 Full load hours VLh (offshore)	16	1.2 Years	48	5 Months
Photovoltaics b)				
Poly-silicon, roof installation, 1000 Full load hours VLh (South Germany)	4,0	6 Years	12	2 Years
Poly-silicon, roof installation, 1800 Full load hours VLh (South Europe)	7,0	3.3 Years	21	1.1 Years

a) The cost of fuel transportation is taken into account b) The values refer to the total energy output. The expense for storage power plants, seasonal reserves or conventional load balancing power plants is not taken into account. c) The data for the E-82 come from the manufacturer, but are confirmed by TÜV Rheinland.

ESOEI

ESOEI (or $ESOI_e$) is used when EROEI is below 1. "$ESOI_e$ is the ratio of electrical energy stored over the lifetime of a storage device to the amount of embodied electrical energy required to build the device."

Storage Technology	ESOEI
Lead acid battery	5
Zinc bromide battery	9
Vanadium redox battery	10
NaS battery	20
Lithium ion battery	32
Pumped hydroelectric storage	704
Compressed air energy storage	792

One of the notable outcomes of the Stanford University team's assessment on ESOI, was that if pumped storage was not available, the combination of wind energy and the commonly suggested pairing with battery technology as it presently exists, would not be sufficiently worth the investment, suggesting instead curtailment.

EROEI under Rapid Growth

A related recent concern is energy cannibalism where energy technologies can have a limited growth rate if climate neutrality is demanded. Many energy technologies are capable of replacing significant volumes of fossil fuels and concomitant green house gas emissions. Unfortunately, neither the enormous scale of the current fossil fuel energy system nor the necessary growth rate of these technologies is well understood within the limits imposed by the net energy produced for a growing industry. This technical limitation is known as energy cannibalism and refers to an effect where rapid growth of an entire energy producing or energy efficiency industry creates a need for energy that uses (or cannibalizes) the energy of existing power plants or production plants.

The solar breeder overcomes some of these problems. A solar breeder is a photovoltaic panel manufacturing plant which can be made energy-independent by using energy derived from its own roof using its own panels. Such a plant becomes not only energy self-sufficient but a major supplier of new energy, hence the name solar breeder. Research on the concept was conducted by Centre for Photovoltaic Engineering, University of New South Wales, Australia. The reported investigation establishes certain mathematical relationships for the solar breeder which clearly indicate that a vast amount of net energy is available from such a plant for the indefinite future. The solar module processing plant at Frederick, Maryland was originally planned as such a solar breeder. In 2009 the Sahara Solar Breeder Project was proposed by the *Science Council of Japan* as a cooperation between Japan and Algeria with the highly ambitious goal of creating hundreds of GW of capacity within 30 years. Theoretically breeders of any kind can be developed. In practice, nuclear breeder reactors are the only large scale breeders that have been constructed as of 2014, with the 600 MWe BN-600 and 800 MWe BN-800 reactor, the two largest in operation.

ENERGY CONTENT OF BIOFUEL

The Energy content of biofuel is a description of the chemical energy contained in a given biofuel, measured per unit mass of that fuel, as specific energy, or per unit of volume of the fuel, as energy density. A biofuel is a fuel, produced from living organisms. Biofuels include bioethanol, an alcohol made by fermentation—often used as a gasoline additive, and biodiesel, which is usually used as a diesel additive. Specific energy is energy per unit mass, which is used to describe the energy content of a fuel, expressed in SI units as joule per kilogram (J/kg) or equivalent units. Energy density is the amount of energy stored in a fuel per unit volume, expressed in SI units as joule per litre (J/L) or equivalent units.

Energy and CO_2 Output of Common Biofuels

The table below includes entries for popular substances already used for their energy, or being discussed for such use.

The second column shows specific energy, the energy content in megajoules per unit of mass in kilograms, useful in understanding the energy that can be extracted from the fuel.

The third column in the table lists energy density, the energy content per liter of volume, which is useful for understanding the space needed for storing the fuel.

The final two columns deal with the carbon footprint of the fuel. The fourth column contains the proportion of CO_2 released when the fuel is converted for energy, with respect to its starting mass, and the fifth column lists the energy produced per kilogram of CO_2 produced. As a guideline, a higher number in this column is better for the environment. But these numbers do not account for other green house gases released during burning, production, storage, or shipping. For example, methane may have hidden environmental costs that are not reflected in the table.

Fuel Type	Specific energy (MJ/kg)	Energy Density (MJ/L)	CO_2 Gas made from Fuel Used (kg/kg)	Energy per CO_2 (MJ/kg)
Solid Fuels				
Bagasse (Cane Stalks)	9.6		~+40%$(C_6H_{10}O_5)_n$+15%$(C_{26}H_{42}O_{21})_n$+15%$(C_9H_{10}O_2)_n$1.30	7.41
Chaff (Seed Casings)	14.6		-	
Animal Dung/ Manure	10-15		-	
Dried plants $(C_6H_{10}O_5)_n$	10 – 16	1.6 - 16.64	IF50%$(C_6H_{10}O_5)_n$+25%$(C_{26}H_{42}O_{21})_n$+25%$(C_{10}H_{12}O_3)_n$1.84	5.44-8.70
Wood fuel $(C_6H_{10}O_5)_n$	16 – 21	2.56 - 21.84	IF45%$(C_6H_{10}O_5)_n$+25%$(C_{26}H_{42}O_{21})_n$+30%$(C_{10}H_{12}O_3)_n$1.88	8.51-11.17
Charcoal	30		85-98% Carbon+VOC+Ash 3.63	8.27
Liquid Fuels				
Pyrolysis oil	17.5	21.35	(Assumption Of Fuel: Carbon Content = 23% w/w) 0.84	20.77
Methanol $(CH_3$-OH)	19.9 – 22.7	15.9	1.37	14.49-16.53
Ethanol $(CH_3$-CH_2-OH)	23.4 – 26.8	18.4 - 21.2	1.91	12.25-14.03
Ecalene™	28.4	22.7	75%C_2H_6O+9%C_3H_8O+7%$C_4H_{10}O$+5%$C_5H_{12}O$+4%Hx 2.03	14.02
Butanol $(CH_3$-$(CH_2)_3$-OH)	36	29.2	2.37	15.16
Fat	37.656	31.68	-	
Biodiesel	37.8	33.3 – 35.7	~2.85	~13.26
Sunflower oil $(C_{18}H_{32}O_2)$	39.49	33.18	(12%$(C_{16}H_{32}O_2)$+16%$(C_{18}H_{34}O_2)$+71%(LA)+1%(ALA))2.81	14.04
Castor oil $(C_{18}H_{34}O_3)$	39.5	33.21	(1%PA+1%SA+89.5%ROA+3%OA+4.2%LA+0.3%ALA)2.67	14.80
Olive oil $(C_{18}H_{34}O_2)$	39.25 - 39.82	33 - 33.48	(15%$(C_{16}H_{32}O_2)$+75%$(C_{18}H_{34}O_2)$+9%(LA)+1%(ALA))2.80	14.03
Gaseous Fuels				
Methane (CH_4)	55 – 55.7	(Liquefied) 23.0 – 23.3	(Methane leak exerts 23 × greenhouse effect of CO_2) 2.74	20.05-20.30
Hydrogen (H_2)	120 – 142	(Liquefied) 8.5 – 10.1	(Hydrogen leak slightly catalyzes ozone depletion) 0.0	

Fossil Fuels (comparison)				
Coal	$29.3 - 33.5$	$39.85 - 74.43$	(Not Counting:CO, NO_x, Sulfates & Particulates) ~3.59	~8.16-9.33
Crude Oil	41.868	$28 - 31.4$	(Not Counting:CO,NO_x,Sulfates & Particulates) ~3.4	~12.31
Gasoline	$45 - 48.3$	$32 - 34.8$	(Not Counting:CO,NO_x,Sulfates & Particulates) ~3.30	~13.64-14.64
Diesel	48.1	40.3	(Not Counting:CO,NO_x,Sulfates & Particulates) ~3.4	~14.15
Natural Gas	$38 - 50$	(Liquefied) $25.5 - 28.7$	(Ethane, Propane & Butane N/C:CO,NO_x & Sulfates) ~3.00	~12.67-16.67
Ethane (CH_3-CH_3)	51.9	(Liquefied) ~24.0	2.93	17.71
Nuclear fuels (comparison)				
Uranium-235 (^{235}U)	77,000,000	(Pure)1,470,700,000	[Greater for lower ore conc. (Mining, Refining, Moving)] 0.0	~55 - ~90
Nuclear fusion (^2H-^3H)	300,000,000	(Liquefied)53,414,377.6	(Sea-Bed Hydrogen-Isotope Mining-Method Dependent) 0.0	
Fuel Cell Energy Storage (comparison)				
Direct-Methanol	4.5466	3.6	~1.37	~3.31
Proton-Exchange (R&D)	up to 5.68	up to 4.5	(IFF Fuel is recycled) 0.0	
Sodium Hydride (R&D)	up to 11.13	up to 10.24	(Bladder for Sodium Oxide Recycling) 0.0	
Battery Energy Storage (comparison)				
Lead-acid battery	0.108	~0.1	(200-600 Deep-Cycle Tolerance) 0.0	
Nickel-iron battery	0.0487 - 0.1127	0.0658 - 0.1772	(<40y Life)(2k-3k Cycle Tolerance IF no Memory effect) 0.0	
Nickel-cadmium battery	0.162 - 0.288	~0.24	(1k-1.5k Cycle Tolerance IF no Memory effect) 0.0	
Nickel metal hydride	0.22 - 0.324	0.36	(300-500 Cycle Tolerance IF no Memory effect) 0.0	
Super iron battery	0.33	(1.5 * NiMH) 0.54	(~300 Deep-Cycle Tolerance) 0.0	
Zinc-air battery	0.396 - 0.72	0.5924 - 0.8442	(Recyclable by Smelting & Remixing, not Recharging) 0.0	
Lithium ion battery	0.54 - 0.72	0.9 - 1.9	(3-5 y Life) (500-1k Deep-Cycle Tolerance) 0.0	
Lithium-Ion -Polymer	0.65 - 0.87	(1.2 * Li-Ion)1.08 - 2.28	(3-5 y Life) (300-500 Deep-Cycle Tolerance) 0.0	
Lithium iron phosphate battery				
DURACELL Zinc-Air	1.0584 - 1.5912	5.148 - 6.3216	(1-3 y Shelf-life) (Recyclable not Rechargeable) 0.0	

| Aluminium battery | 1.8 - 4.788 | 7.56 | (10-30 y Life) (3k+ Deep-Cycle Tolerance) 0.0 | |
| PolyPlusBC Li-Aircell | 3.6 - 32.4 | 3.6 - 17.64 | (May be Rechargeable)(Might leak sulfates) 0.0 | |

1. While all CO_2 gas output ratios are calculated to within a less than 1% margin of error(assuming total oxidation of the carbon content of fuel), ratios preceded by a Tilde (~) indicate a margin of error of up to (but no greater than) 9%. Ratios listed do not include emissions from fuel plant cultivation/Mining, purification/refining and transportation. Fuel availability is typically 74--84.3% NET from source Energy Balance.

2. While Uranium-235 (235U) fission produces no CO_2 gas directly, the indirect fossil fuel burning processes of Mining, Milling, Refining, Moving & Radioactive waste disposal, etc. of intermediate to low-grade uranium ore concentrations produces some amount of carbon dioxide. Studies vary as to how much carbon dioxide is emitted. The United Nations Intergovernmental Panel on Climate Change reports that nuclear produces approximately 40 g of CO_2 per kilowatt hour (11 g/MJ, equivalent to 90 MJ/kg CO_2e). A meta-analysis of a number of studies of nuclear CO_2 lifecycle emissions by academic Benjamin K. Sovacool finds nuclear on average produces 66 g of CO_2 per kilowatt hour (18.3 g/MJ, equivalent to 55 MJ/kg CO_2e). One Australian professor claims that nuclear power produces the equivalent CO_2 gas emissions per MJ of net-output-energy of a Natural Gas fired power station.

Yields of Common Crops Associated with Biofuels Production

Crop	Oil (kg/ha)	Oil (L/ha)	Oil (lb/acre)	Oil (US gal/acre)	Oil per seeds[nc 1] (kg/100 kg)	Melting Range (°C) Oil / Fat	Methyl Ester	Ethyl Ester	Iodine number	Cetane number
Groundnut					(Kernel)42					
Copra					62					
Tallow						35 - 42	16	12	40 - 60	75
Lard						32 - 36	14	10	60 - 70	65
Corn (maize)	145	172	129	18		-5	-10	-12	115 - 124	53
Cashew nut	148	176	132	19						
Oats	183	217	163	23						
Lupine	195	232	175	25						
Kenaf	230	273	205	29						
Calendula	256	305	229	33						
Cotton	273	325	244	35	(Seed)13	-1 - 0	-5	-8	100 - 115	55
Hemp	305	363	272	39						
Soybean	375	446	335	48	14	-16 - -12	-10	-12	125 - 140	53
Coffee	386	459	345	49						
Linseed (flax)	402	478	359	51		-24			178	
Hazelnuts	405	482	362	51						
Euphorbia	440	524	393	56						

Crop	Oil (kg/ha)	Oil (L/ha)	Oil (lb/acre)	Oil (US gal/acre)	Oil per seeds (kg/100 kg)	Melting Range (°C) Oil / Fat	Methyl Ester	Ethyl Ester	Iodine number	Cetane number
Pumpkin seed	449	534	401	57						
Coriander	450	536	402	57						
Mustard seed	481	572	430	61	35					
Camelina	490	583	438	62						
Sesame	585	696	522	74	50					
Safflower	655	779	585	83						
Rice	696	828	622	88						
Tung oil tree	790	940	705	100		-2.5			168	
Sunflowers	800	952	714	102	32	-18 - -17	-12	-14	125 - 135	52
Cocoa (cacao)	863	1,026	771	110						
Peanuts	890	1,059	795	113		3			93	
Opium poppy	978	1,163	873	124						
Rapeseed	1,000	1,190	893	127	37	-10 - 5	-10 - 0	-12 - -2	97 - 115	55 - 58
Olives	1,019	1,212	910	129		-12 - -6	-6	-8	77 - 94	60
Castor beans	1,188	1,413	1,061	151	(Seed)50	-18			85	
Pecan nuts	1,505	1,791	1,344	191						
Jojoba	1,528	1,818	1,365	194						
Jatropha	1,590	1,892	1,420	202						
Macadamia nuts	1,887	2,246	1,685	240						
Brazil nuts	2,010	2,392	1,795	255						
Avocado	2,217	2,638	1,980	282						
Coconut	2,260	2,689	2,018	287		20 - 25	-9	-6	8 - 10	70
Chinese Tallow		4,700		500						
Oil palm	5,000	5,950	4,465	635	20-(Kernal)36	20 - 40	-8 - 21	-8 - 18	12 - 95	65 - 85
Algae		95,000		10,000]						

1. Typical oil extraction from 100 kg of oil seeds.

2. Chinese Tallow (Sapium sebiferum, or Tradica Sebifera) is also known as the "Popcorn Tree".

CELLULOSIC ETHANOL COMMERCIALIZATION

Cellulosic ethanol commercialization is the process of building an industry out of methods of turning cellulose-containing organic matter into cellulosic ethanol for use as a biofuel. Companies, such as Iogen, POET, DuPont, and Abengoa, are building refineries that can process biomass and turn it into bioethanol. Companies, such as Diversa, Novozymes, and Dyadic, are producing enzymes that could enable a cellulosic ethanol future. The shift from food crop feedstocks to waste residues and native grasses offers significant opportunities for a range of players, from farmers to biotechnology firms, and from project developers to investors.

As of 2013, the first commercial-scale plants to produce cellulosic biofuels have begun operating. Multiple pathways for the conversion of different biofuel feedstocks are being used. In the next few years, the cost data of these technologies operating at commercial scale, and their relative performance, will become available. Lessons learnt will lower the costs of the industrial processes involved.

Cellulosic Ethanol Production

Cellulosic ethanol can be produced from a diverse array of feedstocks, such as wood pulp from trees or any plant matter. Instead of taking the grain from wheat and grinding that down to get starch and gluten, then taking the starch, cellulosic ethanol production involves the use of the whole crop. This approach should increase yields and reduce the carbon footprint because the amount of energy-intensive fertilisers and fungicides will remain the same, for a higher output of usable material.

Commercialization by Country

Australia

Ethtec is building a pilot plant in Harwood, New South Wales, which uses wood residues as a feedstock.

Brazil

GranBio (formerly known as GraalBio) is building a facility projected to produce 82 million litres of cellulosic ethanol per year.

Canada

In Canada, Iogen Corp. is a developer of cellulosic ethanol process technology. Iogen has developed a proprietary process and operates a demonstration-scale plant in Ontario. The facility has been designed and engineered to process 40 tons of wheat straw per day into ethanol using enzymes made in an adjacent enzyme manufacturing facility. In 2004, Iogen began delivering its first shipments of cellulosic ethanol into the marketplace. In the near term, the company intends to commercialize its cellulose ethanol process by licensing its technology broadly through turnkey plant construction partnerships. The company is currently evaluating sites in the United States and Canada for its first commercial-scale plant.

Lignol Innovations has a pilot plant, which uses wood as a feedstock, in Vancouver.

In March 2009, KL Energy Corporation of South Dakota and Prairie Green Renewable Energy of Alberta announced their intention to develop a cellulosic ethanol plant near Hudson Bay, Saskatchewan. The Northeast Saskatchewan Renewable Energy Facility will use KL Energy's modern design and engineering to produce ethanol from wood waste.

China

Cellulosic ethanol production currently exists at "pilot" and "commercial demonstration" scale, including a plant in China engineered by SunOpta Inc. and owned and operated by China Resources Alcohol Corporation that is currently producing cellulosic ethanol from corn stover (stalks and leaves) on a continuous, 24-hour-per-day basis.

Denmark

Inbicon's bioethanol plant in Kalundborg, with the capacity to produce 5.4 million liters (1.4 million gallons) annually, was opened in 2009. Believed to be the world's largest cellulosic ethanol plant as of early 2011, the facility runs on about 30,000 metric tons (33,000 tons) of straw per year and the plant employs about 30 people. The plant also produces 13,000 metric tons of lignin pellets per year, used as fuel at combined-heat-and-power plants, and 11,100 metric tons of C5 molasses which is currently used for biomethane production via anaerobic digestion, and has been tested as a high carbohydrate animal feed supplement and potential bio-based feedstock for production of numerous commodity chemicals including diols, glycols, organic acids, and biopolymer precursors and intermediates.

Since October 2010, an E5 blend of 95% gasoline and 5% cellulosic ethanol blend has been available at 100 filling stations across Denmark. Distributed by Statoil, the Bio95 2G mixture uses ethanol derived from wheat straw collected on Danish fields after harvest and produced by Inbicon (a div. of DONG Energy), using enzyme technology from Novozymes.

Commercial or experimental Cellulosic Ethanol Plants in Denmark (Operational or under construction)				
Company	Location	Feedstock	Yearly amount	Operational
Biogasol	Bornholm	Wheat straw	5 megalitre	2012
Ensted-værket	Aabenraa	Wheat straw	? megalitre	2013
Inbicon owned by Dong Energy	Kalundborg, Zealand	Wheat straw	5.4 megalitre	2009
Maabjerg Energy Concept owned by Dong Energy	Maabjerg	Wheat straw	50-70 megalitre	Canceled 2016

Germany

The biofuel company Butalco has recently signed a research and development contract with Hohenheim University. The Institute of Fermentation Technology within the Department of Food Science and Biotechnology at Hohenheim University has been concerned with questions on the production of bioethanol for almost 30 years. The focus in recent years has been on the improvement of the material, energy and life cycle assessment of the production of ethanol. Special interest to BUTALCO is the use of the newly built pilot plant, which is equipped with a safety class 1 approved fermentation room with 4 x 1.5 m³ fermenters. The concept of the plant allows both starch and lignocellulosic based raw materials to be processed. The collaboration will allow BUTALCO to optimise its C5 sugar fermenting and butanol producing yeast strains on a technical scale and produce first amounts of bioethanol from lignocellulose. The whole process of the production of biofuel from the choice of cellulosic biomass feedstock to the conversion into sugars and fermentation through to the purification will be optimised under industrial conditions.

In Straubing, the specialty chemicals company Clariant has been operating a precommercial plant based on its sunliquid process since 2012. The plant is able to produce up to 1000 tons of cellulosic ethanol from agricultural residues such as wheat straw, corn stover or sugarcane bagasse. The process technology uses enzymatic hydrolysis, followed by fermentation of C5 and C6 sugar into ethanol. The company plans to licence the technology worldwide.

India

Cellulosic ethanol production currently exists at "pilot" scale, with efforts being made on utilization of waste lignocellulosic biomass for ethanol production. Pilot scale studies for utilization of pine needles and Lantana weed undertaken at Cellulose and Paper Division, Forest Research Institute, Dehradun, India.

Italy

Italy-based Mossi & Ghisolfi Group broke ground for its 13 million US gallons (49,000 m³) per year cellulosic ethanol facility in Crescentino in northwestern Italy on April 12, 2011. The project will be the largest cellulosic ethanol project in the world, 10 times larger than any of the currently operating demonstration-scale facilities. The plant is "expected to become operational in 2012 and will use a variety of locally sourced feedstocks, beginning with wheat straw and Arundo donax, a perennial giant cane". The company went to bankruptcy on 2018 and had to auction the plant.

Japan

Nippon Oil Corporation and other Japanese manufacturers including Toyota Motor Corporation plan to set up a research body to develop cellulose-derived biofuels. The consortium plans to produce 250,000 kilolitres (1.6 million barrels) per year of bioethanol by March 2014, and produce bioethanol at 40 yen ($0.437) per litre (about $70 a barrel) by 2015.

In March 2009, Honda Motor announced an agreement for the construction of a new cellulosic ethanol research facility in Japan. The new Kazusa-branch facility of the Honda Fundamental Technology Research Center will be built within the Kazusa Akademia Park, in Kisarazu, Chiba. Construction is scheduled to begin in April 2009, with the aim to begin operations in November 2009.

Norway

In October 2010, Norway-based cellulosic ethanol technology developer Weyland commenced production at its 200,000 liter (approximately 53,000 gallon) pilot-scale facility in Bergen, Norway. The plant will demonstrate the company's acid hydrolysis production process, paving the way for a commercial-scale project. The company also plans to market its technology worldwide.

Russia

A commercial factory converting wood (50% softwood + 50% hardwood) into Ethanol is in operation in Northern Russia, the city of Kirov, since 1972 and is still profitable. As side products the company, Kirov Biochemical Works, is offering dry fodder yeast (20 tons/month) and Lignin. To install equipment for drying and burning Lignin, both fresh and accumulated in the landfill, for steam and electricity, a bank loan of $200 million was recently secured.

Spain

Abengoa continues to invest heavily in the necessary technology for bringing cellulosic ethanol to market. Utilizing process and pre-treatment technology from SunOpta Inc., Abengoa is building a 5 million US gallons (19,000 m³) cellulosic ethanol facility in Spain and have recently entered into a strategic research and development agreement with Dyadic International, Inc. (AMEX: DIL), to create new and better enzyme mixtures which may be used to improve both the efficiencies and cost structure of producing cellulosic ethanol.

Sweden

SEKAB has developed an industrial process for production of ethanol from biomass feed-stocks, including wood chips and sugar cane bagasse. The development work is being carried out at an advanced pilot plant in Örnsköldsvik, and has sparked international interest. The technology will be gradually scaled up to commercial production in a new breed of bio-refineries from 2013 to 2015.

United States

The US government actively supports the development and commercialization of cellulosic ethanol through a variety of mechanisms. In the first decade of the 21st century, a lot of companies announced plans to build commercial cellulosic ethanol plants, but most of those plans eventually fell apart, and many of the small companies went bankrupt. Currently (2016), here are many demonstration plants throughout the country, and handful of commercial-scale plants which are in operation or close to it. With the market for cellulosic ethanol in the United States projected to continue growing in the coming years, the outlook for this industry is good.

Government Support

The US Federal government is actively promoting the development of ethanol from cellulosic feed-stocks as an alternative to conventional petroleum transportation fuels. For example, programs sponsored by U.S. Department of Energy (DOE) include research to develop better cellulose hydrolysis enzymes and ethanol-fermenting organisms, to engineering studies of potential processes, to co-funding initial ethanol from cellulosic biomass demonstration and production facilities. This research is conducted by various national laboratories, including the National Renewable Energy Laboratory (NREL), Oak Ridge National Laboratory (ORNL) and Idaho National Laboratory (INL), as well as by universities and private industry. Engineering and construction companies and operating companies are generally conducting the engineering work.

In May 2008, Congress passed a new farm bill that will accelerate the commercialization of advanced biofuels, including cellulosic ethanol. The *Food, Conservation, and Energy Act of 2008* provides for grants covering up to 30% of the cost of developing and building demonstration-scale biorefineries for producing "advanced biofuels," which essentially includes all fuels that are not produced from corn kernel starch. It also allows for loan guarantees of up to $250 million for building commercial-scale biorefineries to produce advanced biofuels.

Using a newly developed tool known as the "Biofuels Deployment Model", Sandia researchers have determined that 21 billion US gallons (79,000,000 m³) of cellulosic ethanol could be produced per year by 2022 without displacing current crops. The Renewable Fuels Standard, part of the 2007 Energy Independence and Security Act, calls for an increase in biofuels production to 36 billion US gallons (140,000,000 m³) a year by 2022.

In January 2011, the USDA approved $405 million in loan guarantees through the 2008 Farm Bill to support the commercialization of cellulosic ethanol at three facilities owned by Coskata, Enerkem and INEOS New Planet BioEnergy. The projects represent a combined 73 million US gallons (280,000 m³) per year production capacity and will begin producing cellulosic ethanol in 2012. The USDA also released a list of advanced biofuel producers who will receive payments to expand the production of advanced biofuels. In July 2011, the US Department of Energy gave in $105 million in loan guarantees to POET for a commercial-scale plant to be built Emmetsburg, Iowa.

Commercial Development

The cellulosic ethanol industry in the United States developed some new commercial-scale plants in 2008. Plants totaling 12 million liters (3.17 million gal) per year were operational, and an additional 80 million liters (21.13 million gal.) per year of capacity - in 26 new plants - was under construction. (For comparison the estimated US petroleum consumption for all uses was about 816 million gal/day in 2008.)

Cellulosic Ethanol Plants in the U.S. (Operational or under construction)					
Company	Location	Feedstock	Capacity (million gal/year)	Operated	Type
Abengoa Bioenergy	Hugoton, KS	Wheat straw	25 - 30	2013 - 2016 (bankrupt)	Commercial
American Process, Inc	Alpena, MI	Wood chips	1.0	2012 -	Demonstration
BlueFire Ethanol	Fulton, MS	Multiple sources	19	Construction halted 2011	Commercial
Coskata, Inc.	Madison, Pennsylvania	Multiple sources	0.04	2009 - 2015	Semi-commercial
DuPont	Nevada, IA	Corn stover	30	2015 - 2017 (shuttered)	Semi-Commercial
Fulcrum BioEnergy	Reno, NV	Municipal solid waste	10	est. end of 2013	Commercial
Gulf Coast Energy	Livingston, AL	Wood waste	0.3	before 2008	Demonstration
KL Energy Corp.	Upton, WY	Wood waste			
Mascoma	Kinross, MI	Wood waste	20		Commercial
POET LLC	Emmetsburg, IA	Corn stover	20 - 25	Sept. 2014	Commercial
POET LLC	Scotland, SD	Corn stover	0.03	2008	Pilot
Xyleco	Woburn, Massachusetts				Demonstration

Environmental Issues

Cellulosic ethanol and grain-based ethanol are, in fact, the same product, but many scientists believe cellulosic ethanol production has distinct environmental advantages over grain-based ethanol production. On a life-cycle basis, ethanol produced from agricultural residues or dedicated cellulosic crops has significantly lower greenhouse gas emissions and a higher sustainability rating than ethanol produced from grain.

According to US Department of Energy studies conducted by the Argonne National Laboratory of the University of Chicago, cellulosic ethanol reduces greenhouse gas emissions (GHG) by 85% over reformulated gasoline. By contrast, starch ethanol (e.g., from corn), which usually uses natural gas to provide energy for the process, reduces greenhouse gas emissions by 18% to 29% over gasoline.

Critics such as Cornell University professor of ecology and agriculture David Pimentel and University of California at Berkeley engineer Tad Patzek question the likelihood of environmental, energy, or economic benefits from cellulosic ethanol technology from non-waste.

References

- What-are-energy-crops: basmati.com, Retrieved 1 May, 2019

- Hall, CA; Lambert, JG; Balogh, SB (2013). "EROI of different fuels and the implications for society". Energy Policy. 64: 141–52. Doi:10.1016/j.enpol.2013.05.049

- Energy-Crops-Introduction: researchgate.net, Retrieved 17 March, 2019

- Atlason, R; Unnthorsson, R (2014). "Ideal EROI (energy return on investment) deepens the understanding of energy systems". Energy. 67: 241–45. Doi:10.1016/j.energy.2014.01.096

- Homer-Dixon, Thomas (2007). The Upside of Down; Catastrophe, Creativity and the Renewal of Civilisation. Island Press. ISBN 978-1-59726-630-7

- Decker, Jeff. Going Against the Grain: Ethanol from Lignocellulosics, Renewable Energy World, January 22, 2009. Retrieved on February 1, 2009

- Kenneth E. Heselton (2004), "Boiler Operator's Handbook". Fairmont Press, 405 pages. ISBN 0881734357

- "Dupont breaks ground at 30 mmgy cellulosic ethanol facility". Ethanol Producer Magazine. November 30, 2012. Retrieved December 15, 2012

Sources of Bioenergy

Some of the most common sources of bioenergy are biomass, natural gas, biogas, biochar, bi-oliquids, renewable natural gas, bagasse, wood gas, biofuel and wood fuel. The chapter closely examines these fundamental sources of bioenergy to provide an extensive understanding of the subject.

BIOMASS

Biomass, as a renewable energy source, refers to biological material that can be used as fuel or for industrial production. It includes plant materials and metabolic wastes from animals and microbes. More specifically, it includes agricultural wastes such as straw, corn stalks, sugarcane leavings, seed hulls, nutshells, and the manure of farm animals. It also includes yard waste, wood, bark, and sawdust. Although fossil fuels (such as coal and petroleum) have their origin in ancient biomass, they are not considered biomass by the generally accepted definition because the original material has been substantially transformed by geological processes.

In this context, biomass may be burned to generate heat and electricity, or it may be used as raw material for the production of biofuels and a variety of chemical substances. Biomass is biodegradable and renewable. The production of biomass is a growing industry, as there is increasing interest in sustainable fuel sources.

Industrial Production

Industrial biomass can be grown from numerous types of plants, including miscanthus, switchgrass, hemp, corn, poplar, willow, sorghum, and sugarcane. It can also be obtained from a variety of tree species, ranging from eucalyptus to oil palm (palm oil). The particular plant used is usually not very important for the end products, but it does affect processing of the raw material.

Plastics from biomass, like some recently developed to dissolve in seawater, are made the same way as petroleum-based plastics, are actually cheaper to manufacture and meet or exceed most performance standards. However, they lack the same water resistance or longevity as conventional plastics.

Environmental Impact

Biomass is part of the carbon cycle. Carbon from the atmosphere is converted into biological matter by photosynthesis. On death or combustion of the material, the carbon goes back into the atmosphere as carbon dioxide (CO_2). This happens over a relatively short timescale and plant matter used as a fuel can be constantly replaced by planting for new growth. Therefore, a reasonably stable level of atmospheric carbon results from its use as a fuel. It is accepted that the amount of carbon stored in dry wood is approximately 50 percent by weight.

Though biomass is a renewable fuel, and is sometimes called a "carbon neutral" fuel, its use can still contribute to global warming. This happens when the natural carbon equilibrium is disturbed; for example by deforestation or urbanization of green sites. When biomass is used as a fuel, as a replacement for fossil fuels, it releases the same amount of CO_2 into the atmosphere. However, when biomass is used for energy production, it is widely considered carbon neutral, or a net reducer of greenhouse gases because of the offset of methane that would have otherwise entered the atmosphere. The carbon in biomass material, which makes up approximately fifty percent of its dry-matter content, is already part of the atmospheric carbon cycle. Biomass absorbs CO_2 from the atmosphere during its growth, after which its carbon reverts to the atmosphere as a mixture of CO_2 and methane (CH_4), depending on the ultimate fate of the biomass material. CH_4 converts to CO_2 in the atmosphere, completing the cycle. In contrast to biomass carbon, the carbon in fossil fuels is taken out of long-term storage and added to the stock of carbon in the atmosphere.

Energy produced from biomass residues displaces the production of an equivalent amount of energy from fossil fuels, leaving the fossil carbon in storage. It also shifts the composition of the recycled carbon emissions associated with the disposal of the biomass residues from a mixture of CO_2 and CH_4, to almost exclusively CO_2. In the absence of energy production applications, biomass residue carbon would be recycled to the atmosphere through some combination of rotting (biodegradation) and open burning. Rotting produces a mixture of up to fifty percent CH_4, while open burning produces five to ten percent CH_4. Controlled combustion in a power plant converts virtually all of the carbon in the biomass to CO_2. Because CH_4 is a much stronger greenhouse gas than CO_2, shifting CH_4 emissions to CO_2 by converting biomass residues to energy significantly reduces the greenhouse warming potential of the recycled carbon associated with other fates or disposal of the biomass residues.

The existing commercial biomass power generating industry in the United States, which consists of approximately 1,700 MW (megawatts) of operating capacity actively supplying power to the grid, produces about 0.5 percent of the U.S. electricity supply. This level of biomass power generation avoids approximately 11 million tons per year of CO_2 emissions from fossil fuel combustion. It also avoids approximately two million tons per year of CH_4 emissions from the biomass residues that, in the absence of energy production, would otherwise be disposed of by burial (in landfills, in disposal piles, or by the plowing under of agricultural residues), by spreading, and by open burning. Biomass power production is at least five times more effective in reducing greenhouse gas emissions than any other greenhouse-gas-neutral power-production technology, such as other renewable and nuclear energy technologies.

In many cases, especially in Europe where huge agricultural developments such as those in the U.S. are not usual, the cost for transporting the biomass exceeds its actual value and therefore the gathering ground has to be limited to a certain small area. This fact leads to only small possible power outputs, around 1 MWel. To set up an economically feasible operation, those power plants have to be equipped with special (ORC) technology, a cycle similar to the water steam power process just with an organic working medium. Such small power plants can be found in Europe.

Despite harvesting, biomass crops may sequester (trap) carbon. For example, soil organic carbon has been observed to be greater in switchgrass stands than in cultivated cropland soil, especially at depths below 12 inches. The grass sequesters the carbon in its increased root biomass. But the perennial grass may need to be allowed to grow for several years before increases are measurable.

Converting Biomass to Energy

Solid biomass, such as wood and garbage, can be burned directly to produce heat. Biomass can also be converted into a gas called biogas or into liquid biofuels such as ethanol and biodiesel. These fuels can then be burned for energy.

Biogas forms when paper, food scraps, and yard waste decompose in landfills, and it can be produced by processing sewage and animal manure in special vessels called digesters.

Ethanol is made from crops such as corn and sugar cane that are fermented to produce fuel ethanol for use in vehicles. Biodiesel is produced from vegetable oils and animal fats and can be used in vehicles and as heating oil.

Biomass as Fuel

Biomass fuels provided about 5% of total primary energy use in the United States in 2018. Of that 5%, about 46% was from biofuels (mainly ethanol), 44% was from wood and wood-derived biomass, and 10% was from the biomass in municipal waste. Researchers are trying to develop ways to use more biomass for fuel.

Advantages of Biomass

As countries struggle to maintain the clean air targets set in the 2015 Paris Agreement, consumers and energy companies are increasingly seeking green alternatives to traditional energy production methods, including biofuels. According to Carbon Brief, since the last European directive on renewables in 2008, bioenergy has provided around half of the growth in the renewables sector.

Final energy from biomass is about 50 EJ of energy – equivalent to 14% of the world's final energy use – though the potential for final energy from biomass worldwide is far higher, and could be as much as 150 EJ by 2035.

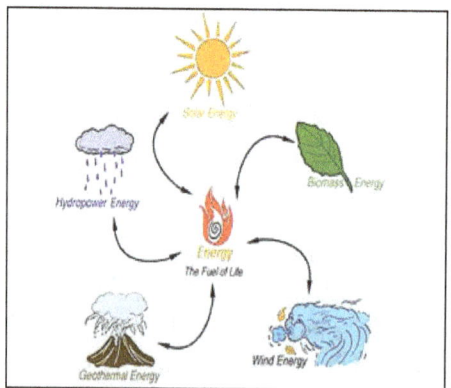

Biomass is an incredibly versatile substance, able to produce energy through being burned directly, converted into liquid biofuels or harvested as a gas from landfills or anaerobic digesters. Its own source of energy comes from the sun, and as plant matter can be regrown relatively quickly, it is classed as renewable.

While a number of waste materials can be used to create biomass – such as sawdust from lumber mills, crop residue, and even chicken litter – it is predominantly sourced from wood. This practice is deemed sustainable by biomass advocates as it can utilise by-products of forest management or help to clear dead or sick trees from an area.

The US Forest Service Wood calls wood an "abundant, sustainable, homegrown cellulosic resource", adding that it could "significantly contribute to meeting 30% of US petroleum consumption from biomass sources by 2030". In 2016, biomass made up 4.8% of total US energy consumption and 12% of renewable energy in the country.

However, critics warn that an over-dependence on a process involving burning trees could have a dangerous effect.

Disadvantages of Biomass

Professor John Beddington, wrote that it "may even lead to a situation whereby global emissions [of carbon dioxide] accelerate", saying we should instead focus our attentions on developing the less damaging sources of wind and solar.

He warns that encouraging the use of bioenergy may lead people to harvest trees and plants specifically for use in power plants rather than merely using the waste products, a fear which will become increasingly likely as demand for biomass grows.

In addition, he notes that burning biomass is not even an entirely clean process, saying wood's lower burning temperature combined with its greater carbon intensity "means wood releases more carbon than fossil fuels per unit of energy generated (almost 4 times more than natural gas, and over 1.5 times that of coal)".

The Earth Institute similarly found current biomass-burning power plants produce 65% more carbon dioxide per megawatt hour than fossil fuel plants, saying it also contributes large amounts of air pollution such as nitrogen oxides, carbon monoxide and lead, many of which are carcinogenic.

As such, critics say the use of biomass should not be welcomed wholesale as a carbon-neutral alternative to coal but instead be approached as a source to be developed with caution, with more transparency about any potentially negative environmental impact essential for consumers.

Biomass Resources

Biomass resources that are available on a renewable basis and are used either directly as a fuel or converted to another form or energy product are commonly referred to as "feedstocks."

Biomass Feedstocks

Biomass feedstocks include dedicated energy crops, agricultural crop residues, forestry residues, algae, wood processing residues, municipal waste, and wet waste (crop wastes, forest residues, purpose-grown grasses, woody energy crops, algae, industrial wastes, sorted municipal solid waste [MSW], urban wood waste, and food waste).

Dedicated Energy Crops

Dedicated energy crops are non-food crops that can be grown on marginal land (land not suitable for traditional crops like corn and soybeans) specifically to provide biomass. These break down into two general categories: herbaceous and woody. Herbaceous energy crops are perennial (plants that live for more than 2 years) grasses that are harvested annually after taking 2 to 3 years to reach full productivity. These include switchgrass, miscanthus, bamboo, sweet sorghum, tall fescue, kochia, wheatgrass, and others. Short-rotation woody crops are fast-growing hardwood trees that are harvested within 5 to 8 years of planting. These include hybrid poplar, hybrid willow, silver maple, eastern cottonwood, green ash, black walnut, sweetgum, and sycamore. Many of these species can help improve water and soil quality, improve wildlife habitat relative to annual crops, diversify sources of income, and improve overall farm productivity.

Agricultural Crop Residue

There are many opportunities to leverage agricultural resources on existing lands without interfering with the production of food, feed, fiber, or forest products. Agricultural crop residues, which include the stalks and leaves, are abundant, diverse, and widely distributed across the United States. Examples include corn stover (stalks, leaves, husks, and cobs), wheat straw, oat straw, barley straw, sorghum stubble, and rice straw. The sale of these residues to a local biorefinery also represents an opportunity for farmers to generate additional income.

Forestry Residues

Forest biomass feedstocks fall into one of two categories: forest residues left after logging timber (including limbs, tops, and culled trees and tree components that would be otherwise unmerchantable) or whole-tree biomass harvested explicitly for biomass. Dead, diseased, poorly formed, and other unmerchantable trees are often left in the woods following timber harvest. This woody debris can be collected for use in bioenergy, while leaving enough behind to provide habitat and maintain proper nutrient and hydrologic features. There are also opportunities to make use of excess biomass on millions of acres of forests. Harvesting excessive woody biomass can reduce the risk of fire and pests, as well as aid in forest restoration, productivity, vitality, and resilience. This biomass could be harvested for bioenergy without negatively impacting the health and stability of forest ecological structure and function.

Algae

Algae as feedstocks for bioenergy refers to a diverse group of highly productive organisms that include microalgae, macroalgae (seaweed), and cyanobacteria (formerly called "blue-green algae"). Many use sunlight and nutrients to create biomass, which contains key components—including lipids, proteins, and carbohydrates— that can be converted and upgraded to a variety of biofuels and products. Depending on the strain, algae can grow by using fresh, saline, or brackish water from surface water sources, groundwater, or seawater. Additionally, they can grow in water from second-use sources, such as treated industrial wastewater; municipal, agricultural, or aquaculture wastewater; or produced water generated from oil and gas drilling operations.

Wood Processing Residues

Wood processing yields byproducts and waste streams that are collectively called wood processing residues and have significant energy potential. For example, the processing of wood for products or pulp produces unused sawdust, bark, branches, and leaves/needles. These residues can then be converted into biofuels or bioproducts. Because these residues are already collected at the point of processing, they can be convenient and relatively inexpensive sources of biomass for energy.

Sorted Municipal Waste

MSW resources include mixed commercial and residential garbage, such as yard trimmings, paper and paperboard, plastics, rubber, leather, textiles, and food wastes. MSW for bioenergy also represents an opportunity to reduce residential and commercial waste by diverting significant volumes from landfills to the refinery.

Wet Waste

Wet waste feedstocks include commercial, institutional, and residential food wastes (particularly those currently disposed of in landfills); organic-rich biosolids (i.e., treated sewage sludge from municipal wastewater); manure slurries from concentrated livestock operations; organic wastes from industrial operations; and biogas (the gaseous product of the decomposition of organic matter in the absence of oxygen) derived from any of the above feedstock streams. Transforming these "waste streams" into energy can help create additional revenue for rural economies and solve waste-disposal problems.

NATURAL GAS

Natural gas (also called fossil gas) is a naturally occurring hydrocarbon gas mixture consisting primarily of methane, but commonly including varying amounts of other higher alkanes, and sometimes a small percentage of carbon dioxide, nitrogen, hydrogen sulfide, or helium. It is formed when layers of decomposing plant and animal matter are exposed to intense heat and pressure under the surface of the Earth over millions of years. The energy that the plants originally obtained from the sun is stored in the form of chemical bonds in the gas.

Natural gas is a non-renewable hydrocarbon used as a source of energy for heating, cooking, and electricity generation. It is also used as a fuel for vehicles and as a chemical feedstock in the manufacture of plastics and other commercially important organic chemicals.

Natural gas is a major cause of climate change, both in itself when leaked and also due to the carbon dioxide it produces when burnt.

Natural gas is found in deep underground rock formations or associated with other hydrocarbon reservoirs in coal beds and as methane clathrates. Petroleum is another resource and fossil fuel found in close proximity to and with natural gas. Most natural gas was created over time by two mechanisms: biogenic and thermogenic. Biogenic gas is created by methanogenic organisms in marshes, bogs, landfills, and shallow sediments. Deeper in the earth, at greater temperature and pressure, thermogenic gas is created from buried organic material.

In petroleum production gas is sometimes burnt as flare gas. Before natural gas can be used as a fuel, most, but not all, must be processed to remove impurities, including water, to meet the specifications of marketable natural gas. The by-products of this processing include: ethane, propane, butanes, pentanes, and higher molecular weight hydrocarbons, hydrogen sulfide (which may be converted into pure sulfur), carbon dioxide, water vapor, and sometimes helium and nitrogen.

Natural gas is often informally referred to simply as "gas", especially when compared to other energy sources such as oil or coal. However, it is not to be confused with gasoline, especially in North America, where the term gasoline is often shortened in colloquial usage to *gas*.

Sources

Natural Gas

Natural gas drilling rig.

In the 19th century, natural gas was usually obtained as a by-product of producing oil, since the small, light gas carbon chains came out of solution as the extracted fluids underwent pressure reduction from the reservoir to the surface, similar to uncapping a soft drink bottle where the carbon dioxide effervesces. Unwanted natural gas was a disposal problem in the active oil fields. If there was not a market for natural gas near the wellhead it was prohibitively expensive to pipe to the end user. In the 19th century and early 20th century, unwanted gas was usually burned off at oil fields. Today, unwanted gas (or stranded gas without a market) associated with oil extraction often is returned to the reservoir with 'injection' wells while awaiting a possible future market or to repressurize the formation, which can enhance extraction rates from other wells. In regions with a high natural gas demand (such as the US), pipelines are constructed when it is economically feasible to transport gas from a wellsite to an end consumer.

In addition to transporting gas via pipelines for use in power generation, other end uses for natural gas include export as liquefied natural gas (LNG) or conversion of natural gas into other liquid products via gas to liquids (GTL) technologies. GTL technologies can convert natural gas into liquids products such as gasoline, diesel or jet fuel. A variety of GTL technologies have been developed, including Fischer–Tropsch (F–T), methanol to gasoline (MTG) and syngas to gasoline plus (STG+). F–T produces a synthetic crude that can be further refined into finished products, while MTG can produce synthetic gasoline from natural gas. STG+ can produce drop-in gasoline, diesel, jet fuel and aromatic chemicals directly from natural gas via a single-loop process. In 2011, Royal Dutch Shell's 140,000 barrels (22,000 m³) per day F–T plant went into operation in Qatar.

Natural gas can be "associated" (found in oil fields), or "non-associated" (isolated in natural gas fields), and is also found in coal beds (as coalbed methane). It sometimes contains a significant amount of ethane, propane, butane, and pentane—heavier hydrocarbons removed for commercial use prior to the methane being sold as a consumer fuel or chemical plant feedstock. Non-hydrocarbons such as carbon dioxide, nitrogen, helium (rarely), and hydrogen sulfide must also be removed before the natural gas can be transported.

Natural gas extracted from oil wells is called casinghead gas (whether or not truly produced up the annulus and through a casinghead outlet) or associated gas. The natural gas industry is extracting an increasing quantity of gas from challenging resource types: sour gas, tight gas, shale gas, and coalbed methane.

There is some disagreement on which country has the largest proven gas reserves. Sources that consider that Russia has by far the largest proven reserves include the US CIA (47 600 km³), the US Energy Information Administration (47 800 km³), and OPEC (48 700 km³). However, BP credits Russia with only 32 900 km³, which would place it in second place, slightly behind Iran (33 100 to 33 800 km³, depending on the source). With Gazprom, Russia is frequently the world's largest natural gas extractor. Major proven resources (in cubic kilometers) are world 187 300 (2013), Iran 33 600 (2013), Russia 32 900 (2013), Qatar 25 100 (2013), Turkmenistan 17 500 (2013) and the United States 8500 (2013).

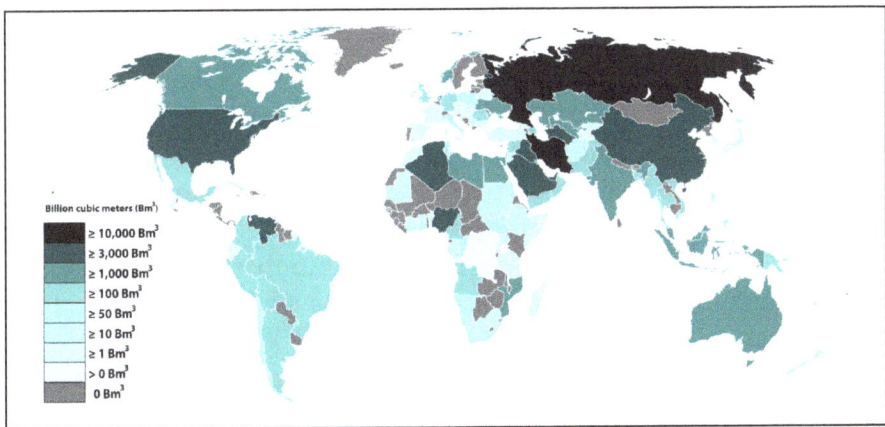

Countries by natural gas proven reserves.

It is estimated that there are about 900 000 km³ of "unconventional" gas such as shale gas, of which 180 000 km³ may be recoverable. In turn, many studies from MIT, Black & Veatch and the DOE predict that natural gas will account for a larger portion of electricity generation and heat in the future.

The world's largest gas field is the offshore South Pars / North Dome Gas-Condensate field, shared between Iran and Qatar. It is estimated to have 51,000 cubic kilometers (12,000 cu mi) of natural gas and 50 billion barrels (7.9 billion cubic meters) of natural gas condensates.

Because natural gas is not a pure product, as the reservoir pressure drops when non-associated gas is extracted from a field under supercritical (pressure/temperature) conditions, the higher molecular weight components may partially condense upon isothermic depressurizing—an effect called retrograde condensation. The liquid thus formed may get trapped as the pores of the gas reservoir get depleted. One method to deal with this problem is to re-inject dried gas free of condensate to maintain the underground pressure and to allow re-evaporation and extraction of condensates. More frequently, the liquid condenses at the surface, and one of the tasks of the gas plant is to collect this condensate. The resulting liquid is called natural gas liquid (NGL) and has commercial value.

Shale Gas

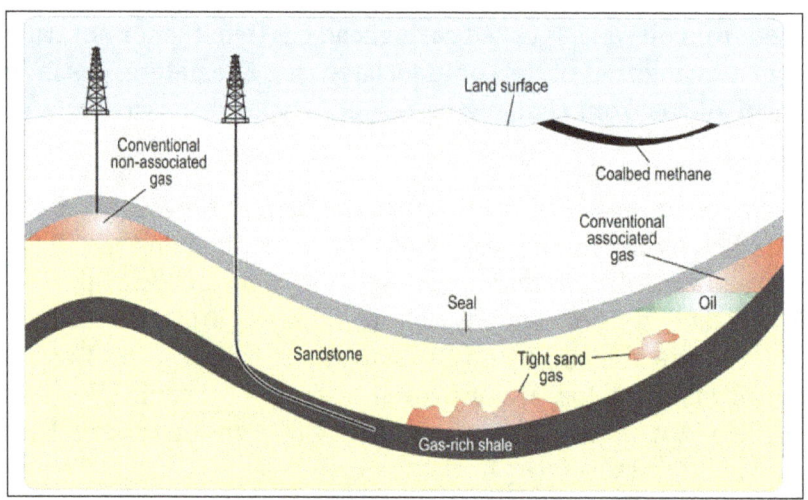

The location of shale gas compared to other types of gas deposits.

Shale gas is natural gas produced from shale. Because shale has matrix permeability too low to allow gas to flow in economical quantities, shale gas wells depend on fractures to allow the gas to flow. Early shale gas wells depended on natural fractures through which gas flowed; almost all shale gas wells today require fractures artificially created by hydraulic fracturing. Since 2000, shale gas has become a major source of natural gas in the United States and Canada. Because of increased shale gas production, the United States is now the number one natural gas producer in the world. Following the success in the United States, shale gas exploration is beginning in countries such as Poland, China, and South Africa.

Town Gas

Town gas is a flammable gaseous fuel made by the destructive distillation of coal. It contains a variety of calorific gases including hydrogen, carbon monoxide, methane, and other volatile hydrocarbons, together with small quantities of non-calorific gases such as carbon dioxide and nitrogen, and is used in a similar way to natural gas. This is a historical technology and is not usually economically competitive with other sources of fuel gas today.

Most town "gashouses" located in the eastern US in the late 19th and early 20th centuries were simple by-product coke ovens that heated bituminous coal in air-tight chambers. The gas driven off from the coal was collected and distributed through networks of pipes to residences and other buildings where it was used for cooking and lighting. (Gas heating did not come into widespread use until the last half of the 20th century.) The coal tar (or asphalt) that collected in the bottoms of the gashouse ovens was often used for roofing and other waterproofing purposes, and when mixed with sand and gravel was used for paving streets.

Biogas

Methanogenic *Archaea* are responsible for almost all biological sources of methane, though methylphosphonate-degrading *Bacteria* produce an as-yet not fully quantified fraction of biogenic methane, particularly in the oceans. Some live in symbiotic relationships with other life forms, including termites, ruminants, and cultivated crops. Other sources of methane, the principal component of natural gas, include landfill gas, biogas, and methane hydrate. When methane-rich gases are produced by the anaerobic decay of organic matter (biomass), these are referred to as biogas (or natural biogas). Sources of biogas include swamps, marshes, and landfills, as well as agricultural waste materials such as sewage sludge and manure by way of anaerobic digesters, in addition to enteric fermentation, particularly in cattle. Landfill gas is created by decomposition of waste in landfill sites. Excluding water vapor, about half of landfill gas is methane and most of the rest is carbon dioxide, with small amounts of nitrogen, oxygen, and hydrogen, and variable trace amounts of hydrogen sulfide and siloxanes. If the gas is not removed, the pressure may get so high that it works its way to the surface, causing damage to the landfill structure, unpleasant odor, vegetation die-off, and an explosion hazard. The gas can be vented to the atmosphere, flared or burned to produce electricity or heat. Biogas can also be produced by separating organic materials from waste that otherwise goes to landfills. This method is more efficient than just capturing the landfill gas it produces. Anaerobic lagoons produce biogas from manure, while biogas reactors can be used for manure or plant parts. Like landfill gas, biogas is mostly methane and carbon dioxide, with small amounts of nitrogen, oxygen and hydrogen. However, with the exception of pesticides, there are usually lower levels of contaminants.

Landfill gas cannot be distributed through utility natural gas pipelines unless it is cleaned up to less than 3% CO_2, and a few parts per million H_2S, because CO_2 and H_2S corrode the pipelines. The presence of CO_2 will lower the energy level of the gas below requirements for the pipeline. Siloxanes in the gas will form deposits in gas burners and need to be removed prior to entry into any gas distribution or transmission system. Consequently, it may be more economical to burn the gas on site or within a short distance of the landfill using a dedicated pipeline. Water vapor is often removed, even if the gas is burned on site. If low temperatures condense water out of the gas, siloxanes can be lowered as well because they tend to condense out with the water vapor. Other non-methane components may also be removed to meet emission standards, to prevent fouling of the equipment or for environmental considerations. Co-firing landfill gas with natural gas improves combustion, which lowers emissions.

Biogas, and especially landfill gas, are already used in some areas, but their use could be greatly expanded. Experimental systems were being proposed for use in parts of Hertfordshire, UK, and

Lyon in France. Using materials that would otherwise generate no income, or even cost money to get rid of, improves the profitability and energy balance of biogas production. Gas generated in sewage treatment plants is commonly used to generate electricity. For example, the Hyperion sewage plant in Los Angeles burns 8 million cubic feet (230,000 cubic meters) of gas per day to generate power New York City utilizes gas to run equipment in the sewage plants, to generate electricity, and in boilers. Using sewage gas to make electricity is not limited to large cities. The city of Bakersfield, California, uses cogeneration at its sewer plants. California has 242 sewage wastewater treatment plants, 74 of which have installed anaerobic digesters. The total biopower generation from the 74 plants is about 66 MW.

Crystallized Natural Gas — Hydrates

Huge quantities of natural gas (primarily methane) exist in the form of hydrates under sediment on offshore continental shelves and on land in arctic regions that experience permafrost, such as those in Siberia. Hydrates require a combination of high pressure and low temperature to form.

In 2010, the cost of extracting natural gas from crystallized natural gas was estimated to be as much as twice the cost of extracting natural gas from conventional sources, and even higher from offshore deposits.

In 2013, Japan Oil, Gas and Metals National Corporation (JOGMEC) announced that they had recovered commercially relevant quantities of natural gas from methane hydrate.

The McMahon natural gas processing plant.

Processing

The image below is a schematic block flow diagram of a typical natural gas processing plant. It shows the various unit processes used to convert raw natural gas into sales gas pipelined to the end user markets.

The block flow diagram also shows how processing of the raw natural gas yields byproduct sulfur, byproduct ethane, and natural gas liquids (NGL) propane, butanes and natural gasoline (denoted as pentanes +).

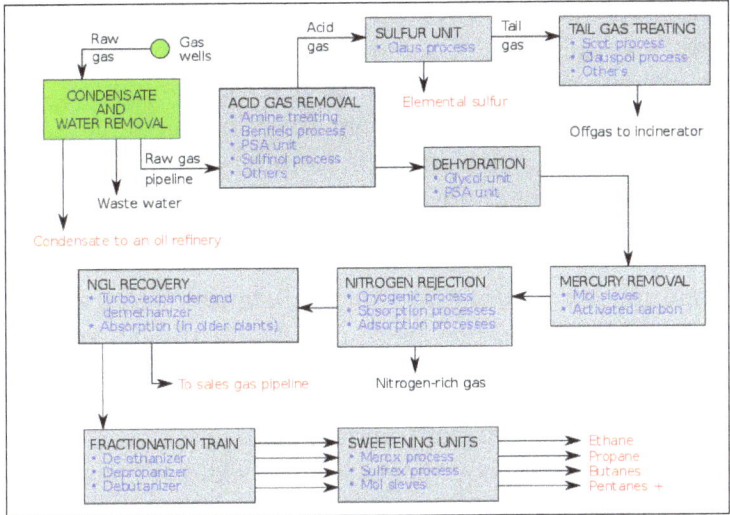

Schematic flow diagram of a typical natural gas processing plant.

Depletion

As of 2018, natural gas production in the US has peaked twice, with current levels exceeding both previous peaks. It reached 24.1 million cubic feet in 1973, followed by a decline, and reached 24.5 million cubic feet in 2001. After a brief drop, withdrawals have been increasing nearly every year since 2006, with 2017 production at 33.4 million cubic feet.

Storage and Transport

Polyethylene plastic main being placed in a trench.

Because of its low density, it is not easy to store natural gas or to transport it by vehicle. Natural gas pipelines are impractical across oceans, since the gas needs to be cooled down and compressed, as the friction in the pipeline causes the gas to heat up. Many existing pipelines in America are close to reaching their capacity, prompting some politicians representing northern states to speak of potential shortages. The large trade cost implies that natural gas markets are globally much less integrated, causing significant price differences across countries. In Western Europe, the gas pipeline network is already dense. New pipelines are planned or under construction in Eastern Europe and between gas fields in Russia, Near East and Northern Africa and Western Europe.

Whenever gas is bought or sold at custody transfer points, rules and agreements are made regarding the gas quality. These may include the maximum allowable concentration of CO_2, H_2S and H_2O. Usually sales quality gas that has been treated to remove contamination is traded on a "dry gas" basis and is required to be commercially free from objectionable odours, materials, and dust or other solid or liquid matter, waxes, gums and gum forming constituents, which might damage or adversely affect operation of equipment downstream of the custody transfer point.

LNG carriers transport liquefied natural gas (LNG) across oceans, while tank trucks can carry liquefied or compressed natural gas (CNG) over shorter distances. Sea transport using CNG carrier ships that are now under development may be competitive with LNG transport in specific conditions.

Gas is turned into liquid at a liquefaction plant, and is returned to gas form at regasification plant at the terminal. Shipborne regasification equipment is also used. LNG is the preferred form for long distance, high volume transportation of natural gas, whereas pipeline is preferred for transport for distances up to 4,000 km (2,500 mi) over land and approximately half that distance offshore.

CNG is transported at high pressure, typically above 200 bars (20,000 kPa; 2,900 psi). Compressors and decompression equipment are less capital intensive and may be economical in smaller unit sizes than liquefaction/regasification plants. Natural gas trucks and carriers may transport natural gas directly to end-users, or to distribution points such as pipelines.

Peoples Gas Manlove Field natural gas storage area. In the foreground (left) is one of the numerous wells for the underground storage area, with an LNG plant, and above ground storage tanks are in the background (right).

In the past, the natural gas which was recovered in the course of recovering petroleum could not be profitably sold, and was simply burned at the oil field in a process known as flaring. Flaring is now illegal in many countries. Additionally, higher demand in the last 20–30 years has made production of gas associated with oil economically viable. As a further option, the gas is now sometimes re-injected into the formation for enhanced oil recovery by pressure maintenance as well as miscible or immiscible flooding. Conservation, re-injection, or flaring of natural gas associated with oil is primarily dependent on proximity to markets (pipelines), and regulatory restrictions.

Natural gas can be indirectly exported through the absorption in other physical output. A recent study suggests that the expansion of shale gas production in the US has caused prices to drop relative to other countries. This has caused a boom in energy intensive manufacturing sector exports, whereby the average dollar unit of US manufacturing exports has almost tripled its energy content between 1996 and 2012.

A "master gas system" was invented in Saudi Arabia in the late 1970s, ending any necessity for flaring. Satellite observation, however, shows that flaring and venting are still practiced in some gas-extracting countries.

Natural gas is used to generate electricity and heat for desalination. Similarly, some landfills that also discharge methane gases have been set up to capture the methane and generate electricity.

Natural gas is often stored underground inside depleted gas reservoirs from previous gas wells, salt domes, or in tanks as liquefied natural gas. The gas is injected in a time of low demand and extracted when demand picks up. Storage nearby end users helps to meet volatile demands, but such storage may not always be practicable.

With 15 countries accounting for 84% of the worldwide extraction, access to natural gas has become an important issue in international politics, and countries vie for control of pipelines. In the first decade of the 21st century, Gazprom, the state-owned energy company in Russia, engaged in disputes with Ukraine and Belarus over the price of natural gas, which have created concerns that gas deliveries to parts of Europe could be cut off for political reasons. The United States is preparing to export natural gas.

Floating Liquefied Natural Gas

Floating liquefied natural gas (FLNG) is an innovative technology designed to enable the development of offshore gas resources that would otherwise remain untapped due to environmental or economic factors it is nonviable to develop them via a land-based LNG operation. FLNG technology also provides a number of environmental and economic advantages:

- Environmental – Because all processing is done at the gas field, there is no requirement for long pipelines to shore, compression units to pump the gas to shore, dredging and jetty construction, and onshore construction of an LNG processing plant, which significantly reduces the environmental footprint. Avoiding construction also helps preserve marine and coastal environments. In addition, environmental disturbance will be minimised during decommissioning because the facility can easily be disconnected and removed before being refurbished and re-deployed elsewhere.

- Economic – Where pumping gas to shore can be prohibitively expensive, FLNG makes development economically viable. As a result, it will open up new business opportunities for countries to develop offshore gas fields that would otherwise remain stranded, such as those offshore East Africa.

Many gas and oil companies are considering the economic and environmental benefits of floating liquefied natural gas (FLNG). There are currently projects underway to construct five FLNG facilities. Petronas is close to completion on their FLNG-1 at Daewoo Shipbuilding and Marine Engineering and are underway on their FLNG-2 project at Samsung Heavy Industries. Shell Prelude is due to start production 2017. The Browse LNG project will commence FEED in 2019.

Uses

Natural gas is primarily used in the northern hemisphere. North America and Europe are major consumers.

Mid-stream Natural Gas

Often well head gases require removal of various hydrocarbon molecules contained within the gas. Some of these gases include heptane, pentane, propane and other hydrocarbons with molecular weights above methane (CH_4). The natural gas transmission lines extend to the natural gas processing plant or unit which removes the higher molecular weighted hydrocarbons to produce natural gas with energy content between 950–1,050 British thermal units per cubic foot (35–39 MJ/m³). The processed natural gas may then be used for residential, commercial and industrial uses.

Natural gas flowing in the distribution lines is called mid-stream natural gas and is often used to power engines which rotate compressors. These compressors are required in the transmission line to pressurize and repressurize the mid-stream natural gas as the gas travels. Typically, natural gas powered engines require 950–1,050 BTU/cu ft (35–39 MJ/m³) natural gas to operate at the rotational name plate specifications. Several methods are used to remove these higher molecular weighted gases for use by the natural gas engine. A few technologies are as follows:

- Joule–Thomson skid
- Cryogenic or chiller system
- Chemical enzymology system

Power Generation

Natural gas is a major source of electricity generation through the use of cogeneration, gas turbines and steam turbines. Natural gas is also well suited for a combined use in association with renewable energy sources such as wind or solar and for alimenting peak-load power stations functioning in tandem with hydroelectric plants. Most grid peaking power plants and some off-grid engine-generators use natural gas. Particularly high efficiencies can be achieved through combining gas turbines with a steam turbine in combined cycle mode. Natural gas burns more cleanly than other fuels, such as oil and coal. Because burning natural gas produces both water and carbon dioxide, it produces less carbon dioxide per unit of energy released than coal, which produces mostly carbon dioxide. Burning natural gas produces only about half the carbon dioxide per kilowatt-hour (kWh) that coal does. For transportation, burning natural gas produces about 30% less carbon dioxide than burning petroleum. The US Energy Information Administration reports the following emissions in million metric tons of carbon dioxide in the world for 2012:

- Natural gas: 6,799
- Petroleum: 11,695
- Coal: 13,787

Coal-fired electric power generation emits around 2,000 pounds (900 kg) of carbon dioxide for every megawatt-hour (MWh) generated, which is almost double the carbon dioxide released by natural gas-fired generation. Because of this higher carbon efficiency of natural gas generation, as the fuel mix in the United States has changed to reduce coal and increase natural gas generation, carbon dioxide emissions have unexpectedly fallen. Those measured in the first quarter of 2012 were the lowest of any recorded for the first quarter of any year since 1992.

Combined cycle power generation using natural gas is currently the cleanest available source of power using hydrocarbon fuels, and this technology is widely and increasingly used as natural gas can be obtained at increasingly reasonable costs. Fuel cell technology may eventually provide cleaner options for converting natural gas into electricity, but as yet it is not price-competitive. Locally produced electricity and heat using natural gas powered Combined Heat and Power plant (CHP or Cogeneration plant) is considered energy efficient and a rapid way to cut carbon emissions.

Natural gas generated power has increased from 740 TWh in 1973 to 5140 TWh in 2014, generating 22% of the worlds total electricity. Approximately half as much as generated with coal. Efforts around the world to reduce the use of coal has led some regions to switch to natural gas.

Domestic use

Natural gas dispensed in a residential setting can generate temperatures in excess of 1,100 °C (2,000 °F) making it a powerful domestic cooking and heating fuel. In much of the developed world it is supplied through pipes to homes, where it is used for many purposes including ranges and ovens, gas-heated clothes dryers, heating/cooling, and central heating. Heaters in homes and other buildings may include boilers, furnaces, and water heaters. Both North America and Europe are major consumers of natural gas.

Domestic appliances, furnaces, and boilers use low pressure, usually 6 to 7 inches of water (6" to 7" WC), which is about 0.25 psig. The pressures in the supply lines vary, either utilization pressure (UP, the aforementioned 6" to 7" WC) or elevated pressure (EP), which may be anywhere from 1 psig to 120 psig. Systems using EP have a regulator at the service entrance to step down the pressure to UP.

In the US compressed natural gas (CNG) is available in some rural areas as an alternative to less expensive and more abundant LPG (liquefied petroleum gas), the dominant source of rural gas. It is used in homes lacking direct connections to public utility provided gas, or to fuel portable grills. Natural gas is also supplied by independent natural gas suppliers through Natural Gas Choice programs throughout the United States.

A Washington, D.C. Metrobus, which runs on natural gas.

Transportation

CNG is a cleaner and also cheaper alternative to other automobile fuels such as gasoline (petrol) and diesel. By the end of 2014 there were over 20 million natural gas vehicles worldwide, led by Iran (3.5 million), China (3.3 million), Pakistan (2.8 million), Argentina (2.5 million), India (1.8 million), and Brazil (1.8 million). The energy efficiency is generally equal to that of gasoline engines, but lower compared with modern diesel engines. Gasoline/petrol vehicles converted to run on natural gas suffer because of the low compression ratio of their engines, resulting in a cropping of delivered power while running on natural gas (10–15%). CNG-specific engines, however, use a higher compression ratio due to this fuel's higher octane number of 120–130.

Besides use in road vehicles, CNG can also be used in aircraft. Compressed natural gas has been used in some aircraft like the Aviat Aircraft Husky 200 CNG and the Chromarat VX-1 KittyHawk.

LNG is also being used in aircraft. Russian aircraft manufacturer Tupolev for instance is running a development program to produce LNG- and hydrogen-powered aircraft. The program has been running since the mid-1970s, and seeks to develop LNG and hydrogen variants of the Tu-204 and Tu-334 passenger aircraft, and also the Tu-330 cargo aircraft. Depending on the current market price for jet fuel and LNG, fuel for an LNG-powered aircraft could cost 5,000 rubles (US$100) less per tonne, roughly 60%, with considerable reductions to carbon monoxide, hydrocarbon and nitrogen oxide emissions.

The advantages of liquid methane as a jet engine fuel are that it has more specific energy than the standard kerosene mixes do and that its low temperature can help cool the air which the engine compresses for greater volumetric efficiency, in effect replacing an intercooler. Alternatively, it can be used to lower the temperature of the exhaust.

Fertilizers

Natural gas is a major feedstock for the production of ammonia, via the Haber process, for use in fertilizer production.

Hydrogen

Natural gas can be used to produce hydrogen, with one common method being the hydrogen reformer. Hydrogen has many applications: it is a primary feedstock for the chemical industry, a hydrogenating agent, an important commodity for oil refineries, and the fuel source in hydrogen vehicles.

Animal and Fish Feed

Protein rich animal and fish feed is produced by feeding natural gas to Methylococcus capsulatus bacteria on commercial scale.

Other

Natural gas is also used in the manufacture of fabrics, glass, steel, plastics, paint, and other products.

Environmental Effects

Effect of Natural Gas Release

Natural gas is mainly composed of methane. After release to the atmosphere it is removed by gradual oxidation to carbon dioxide and water by hydroxyl radicals (OH^-) formed in the troposphere or stratosphere, giving the overall chemical reaction $CH_4 + 2O_2 \rightarrow CO_2 + 2H_2O$. While the lifetime of atmospheric methane is relatively short when compared to carbon dioxide, with a half-life of about 7 years, it is more efficient at trapping heat in the atmosphere, so that a given quantity of methane has 84 times the global-warming potential of carbon dioxide over a 20-year period and 28 times over a 100-year period. Natural gas is thus a more potent greenhouse gas than carbon dioxide due to the greater global-warming potential of methane. 2009 estimates by the EPA place global emissions of methane at 85 cubic kilometers (3.0 trillion cubic feet) annually, or 3% of global production, 3.0 trillion cubic meters or 105 trillion cubic feet (2009 est). Direct emissions of methane represented 14.3% by volume of all global anthropogenic greenhouse gas emissions in 2004.

During extraction, storage, transportation, and distribution, natural gas is known to leak into the atmosphere, particularly during the extraction process. In 2021 MethaneSAT should reduce the large uncertainties in the estimates.

Carbon Dioxide Emissions

Natural gas is often described as the cleanest fossil fuel. It produces 25–30% and 40–45% less carbon dioxide per joule delivered than oil and coal respectively, and potentially fewer pollutants than other hydrocarbon fuels. However, in absolute terms, it comprises a substantial percentage of human carbon emissions, and this contribution is projected to grow. According to the IPCC Fourth Assessment Report, in 2004, natural gas produced about 5.3 billion tons a year of CO_2 emissions, while coal and oil produced 10.6 and 10.2 billion tons respectively. According to an updated version of the Special Report on Emissions Scenario by 2030, natural gas would be the source of 11 billion tons a year, with coal and oil now 8.4 and 17.2 billion respectively because demand is increasing 1.9% a year. According to Global Energy Monitor gas pipelines built in the 2010s, especially in the United States, are locking in huge greenhouse gas emissions for 40 to 50 years at a time.

To reduce its greenhouse emissions, the government of the Netherlands is subsidizing a transition away from natural gas for all homes in the country by 2050. In Amsterdam, no new residential gas accounts are allowed as of July 1, 2018, and all homes in the city are expected to be converted by 2040.

Other Pollutants

Natural gas produces far lower amounts of sulfur dioxide and nitrous oxides than other fossil fuels. The pollutants due to natural gas combustion are listed below:

Comparison of emissions from natural gas, oil and coal burning			
Pollutant (lb/MMBtu)	NG	Oil	Coal
Carbon dioxide	117	164	208
Carbon monoxide	0.040	0.033	0.208

Sulfur dioxide	0.001	1.122	2.591
Nitrogen oxides	0.092	0.448	0.457
Particulates	0.007	0.084	2.744
Mercury	0	0.000007	0.000016

Radionuclides

Natural gas extraction also produces radioactive isotopes of polonium (Po-210), lead (Pb-210) and radon (Rn-220). Radon is a gas with initial activity from 5 to 200,000 becquerels per cubic meter of gas. It decays rapidly to Pb-210 which can build up as a thin film in gas extraction equipment.

Safety Concerns

A pipeline odorant injection station.

The natural gas extraction workforce face unique health and safety challenges and is recognized by the National Institute for Occupational Safety and Health (NIOSH) as a priority industry sector in the National Occupational Research Agenda (NORA) to identify and provide intervention strategies regarding occupational health and safety issues.

Production

Some gas fields yield sour gas containing hydrogen sulfide (H_2S), a toxic compound when inhaled. Amine gas treating, an industrial scale process which removes acidic gaseous components, is often used to remove hydrogen sulfide from natural gas.

Extraction of natural gas (or oil) leads to decrease in pressure in the reservoir. Such decrease in pressure in turn may result in subsidence, sinking of the ground above. Subsidence may affect ecosystems, waterways, sewer and water supply systems, foundations, and so on.

Fracking

Releasing natural gas from subsurface porous rock formations may be accomplished by a process called hydraulic fracturing or "fracking". It's estimated that hydraulic fracturing will eventually account for nearly 70% of natural gas development in North America. Since the first commercial hydraulic fracturing operation in 1949, approximately one million wells have been hydraulically fractured in the United States. The production of natural gas from hydraulically fractured wells has utilized the technological developments of directional and horizontal drilling, which improved access to natural gas in tight rock formations. Strong growth in the production of unconventional gas from hydraulically fractured wells occurred between 2000–2012.

In hydraulic fracturing, well operators force water mixed with a variety of chemicals through the wellbore casing into the rock. The high pressure water breaks up or "fracks" the rock, which releases gas from the rock formation. Sand and other particles are added to the water as a proppant to keep the fractures in the rock open, thus enabling the gas to flow into the casing and then to the surface. Chemicals are added to the fluid to perform such functions as reducing friction and inhibiting corrosion. After the "frack," oil or gas is extracted and 30–70% of the frack fluid, i.e. the mixture of water, chemicals, sand, etc., flows back to the surface. Many gas-bearing formations also contain water, which will flow up the wellbore to the surface along with the gas, in both hydraulically fractured and non-hydraulically fractured wells. This produced water often has a high content of salt and other dissolved minerals that occur in the formation.

The volume of water used to hydraulically fracture wells varies according to the hydraulic fracturing technique. In the United States, the average volume of water used per hydraulic fracture has been reported as nearly 7,375 gallons for vertical oil and gas wells prior to 1953, nearly 197,000 gallons for vertical oil and gas wells between 2000–2010, and nearly 3 million gallons for horizontal gas wells between 2000–2010.

Determining which fracking technique is appropriate for well productivity depends largely on the properties of the reservoir rock from which to extract oil or gas. If the rock is characterized by low-permeability — which refers to its ability to let substances, i.e. gas, pass through it, then the rock may be considered a source of tight gas. Fracking for shale gas, which is currently also known as a source of unconventional gas, involves drilling a borehole vertically until it reaches a lateral shale rock formation, at which point the drill turns to follow the rock for hundreds or thousands of feet horizontally. In contrast, conventional oil and gas sources are characterized by higher rock permeability, which naturally enables the flow of oil or gas into the wellbore with less intensive hydraulic fracturing techniques than the production of tight gas has required. The decades in development of drilling technology for conventional and unconventional oil and gas production has not only improved access to natural gas in low-permeability reservoir rocks, but also posed significant adverse impacts on environmental and public health.

The US EPA has acknowledged that toxic, carcinogenic chemicals, i.e. benzene and ethylbenzene, have been used as gelling agents in water and chemical mixtures for high volume horizontal fracturing (HVHF). Following the hydraulic fracture in HVHF, the water, chemicals, and frack fluid that return to the well's surface, called flowback or produced water, may contain radioactive materials, heavy metals, natural salts, and hydrocarbons which exist naturally in shale rock formations. Fracking chemicals, radioactive materials, heavy metals, and salts that are removed from the HVHF well by well operators are so difficult to remove from the water they're mixed with, and would so heavily pollute the water cycle, that most of the flowback is either recycled into other fracking operations or injected into deep underground wells, eliminating the water that HVHF required from the hydrologic cycle.

Added Odor

Natural gas in its native state is colorless and almost odorless. In order to assist consumers in detecting leaks, an odorizer with a scent similar to rotten eggs, tert-Butylthiol (t-butyl mercaptan), is added. Sometimes a related compound, thiophane, may be used in the mixture. Situations in which an odorant that is added to natural gas can be detected by analytical instrumentation, but cannot be properly detected by an observer with a normal sense of smell, have occurred in the natural gas industry. This is caused by odor masking, when one odorant overpowers the sensation of another. As of 2011, the industry is conducting research on the causes of odor masking.

Risk of Explosion

Gas network emergency vehicle responding to a major fire.

Explosions caused by natural gas leaks occur a few times each year. Individual homes, small businesses and other structures are most frequently affected when an internal leak builds up gas inside the structure. Frequently, the blast is powerful enough to significantly damage a building but leave it standing. In these cases, the people inside tend to have minor to moderate injuries. Occasionally, the gas can collect in high enough quantities to cause a deadly explosion, disintegrating one or more buildings in the process. The gas usually dissipates readily outdoors, but can sometimes collect in dangerous quantities if flow rates are high enough. However, considering the tens of millions of structures that use the fuel, the individual risk of using natural gas is very low.

Risk of Carbon Monoxide Inhalation

Natural gas heating systems may cause carbon monoxide poisoning if unvented or poorly vented. In 2011, natural gas furnaces, space heaters, water heaters and stoves were blamed for 11 carbon monoxide deaths in the US. Another 22 deaths were attributed to appliances running on liquified petroleum gas, and 17 deaths on gas of unspecified type. Improvements in natural gas furnace designs have greatly reduced CO poisoning concerns. Detectors are also available that warn of carbon monoxide and/or explosive gas (methane, propane, etc.).

Energy Content, Statistics and Pricing

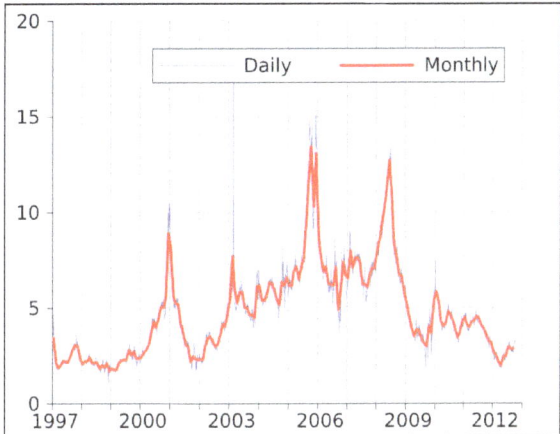

Natural gas prices at the Henry Hub in US dollars per million BTUs ($/mmbtu).

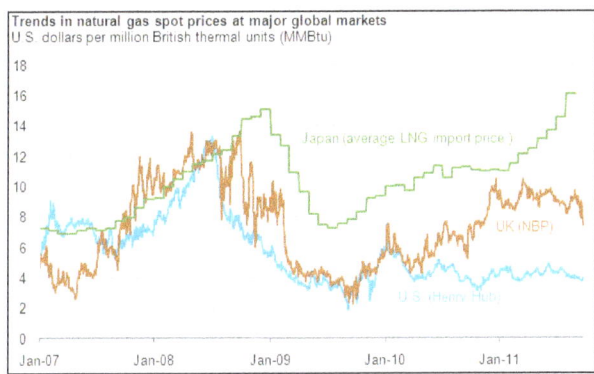

Comparison of natural gas prices in Japan, United Kingdom, and United States, 2007–2011.

Quantities of natural gas are measured in normal cubic meters (cubic meter of gas at "normal" temperature 0 °C (32 °F) and pressure 101.325 kPa (14.6959 psi)) or standard cubic feet (cubic foot of gas at "standard" temperature 60.0 °F (15.6 °C) and pressure 14.73 psi (101.6 kPa)), one cubic meter ≈ 35.3147 cu ft. The gross heat of combustion of commercial quality natural gas is around 39 MJ/m³ (0.31 kWh/cu ft), but this can vary by several percent. This is about 49 MJ/kg (6.2 kWh/lb) (assuming a density of 0.8 kg/m³ (0.05 lb/cu ft), an approximate value).

European Union

Gas prices for end users vary greatly across the EU. A single European energy market, one of the key objectives of the EU, should level the prices of gas in all EU member states. Moreover, it would help to resolve supply and global warming issues, as well as strengthen relations with other Mediterranean countries and foster investments in the region.

United States

In US units, one standard cubic foot (28 L) of natural gas produces around 1,028 British thermal units (1,085 kJ). The actual heating value when the water formed does not condense is the net heat of combustion and can be as much as 10% less.

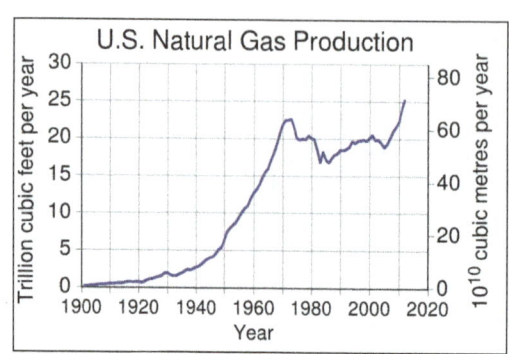

US Natural Gas Marketed Production 1900 to 2012.

In the United States, retail sales are often in units of therms (th); 1 therm = 100,000 BTU. Gas sales to domestic consumers are often in units of 100 standard cubic feet (scf). Gas meters measure the volume of gas used, and this is converted to therms by multiplying the volume by the energy content of the gas used during that period, which varies slightly over time. The typical annual consumption of a single family residence is 1,000 therms or one Residential Customer Equivalent (RCE). Wholesale transactions are generally done in decatherms (Dth), thousand decatherms (MDth), or million decatherms (MMDth). A million decatherms is a trillion BTU, roughly a billion cubic feet of natural gas.

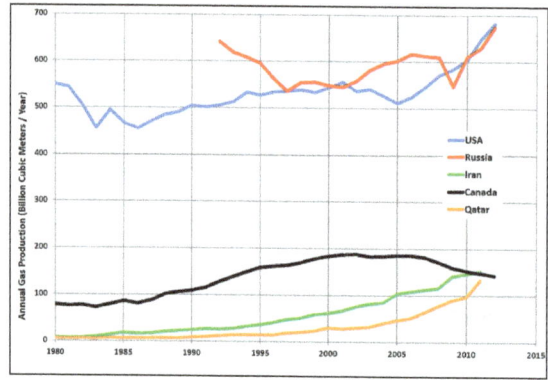

Trends in the top five natural gas-producing countries.

The price of natural gas varies greatly depending on location and type of consumer. In 2007, a price of $7 per 1000 cubic feet ($0.25/m³) was typical in the United States. The typical caloric value of natural gas is roughly 1,000 BTU per cubic foot, depending on gas composition. This corresponds to around $7 per million BTU or around $7 per gigajoule (GJ). In April 2008, the wholesale price was $10 per 1000 cubic feet ($10/MMBTU). The residential price varies from 50% to 300% more than the wholesale price. At the end of 2007, this was $12–$16 per 1000 cubic feet ($0.42–$0.57/m³). Natural gas in the United States is traded as a futures contract on the New York Mercantile Exchange. Each contract is for 10,000 MMBTU or 10 billion BTU (10,551 GJ). Thus, if the price of gas is $10/MMBTU on the NYMEX, the contract is worth $100,000.

Canada

Canada uses metric measure for internal trade of petrochemical products. Consequently, natural gas is sold by the gigajoule (GJ), cubic meter (m³) or thousand cubic meters (E3m3). Distribution infrastructure and meters almost always meter volume (cubic foot or cubic meter). Some jurisdictions, such as Saskatchewan, sell gas by volume only. Other jurisdictions, such as Alberta, gas is sold by the energy content (GJ). In these areas, almost all meters for residential and small commercial customers measure volume (m³ or ft³), and billing statements include a multiplier to convert the volume to energy content of the local gas supply.

A gigajoule (GJ) is a measure approximately equal to half a barrel (250 lbs) of oil, or 1 million BTUs, or 1,000 cu ft or 28 m³ of gas. The energy content of gas supply in Canada can vary from 37 to 43 MJ/m³ (990 to 1,150 BTU/cu ft) depending on gas supply and processing between the wellhead and the customer.

Elsewhere

Outside of the European Union, the U.S., and Canada, natural gas is sold in gigajoule retail units. LNG (liquefied natural gas) and LPG (liquefied petroleum gas) are traded in metric tonnes (1,000 kg) or MMBTU as spot deliveries. Long term natural gas distribution contracts are signed in cubic meters, and LNG contracts are in metric tonnes. The LNG and LPG is transported by specialized transport ships, as the gas is liquified at cryogenic temperatures. The specification of each LNG/LPG cargo will usually contain the energy content, but this information is in general not available to the public.

In the Russian Federation, Gazprom sold approximately 250 billion cubic meters (8.8 trillion cubic feet) of natural gas in 2008. In 2013 they produced 487.4 billion cubic meters (17.21 trillion cubic feet) of natural and associated gas. Gazprom supplied Europe with 161.5 billion cubic meters (5.70 trillion cubic feet) of gas in 2013.

In August 2015, possibly the largest natural gas discovery in history was made and notified by an Italian gas company ENI. The energy company indicated that it has unearthed a "supergiant" gas field in the Mediterranean Sea covering about 40 square miles (100 km²). It was also reported that the gas field could hold a potential 30 trillion cubic feet (850 billion cubic meters) of natural gas. ENI said that the energy is about5.5 billion barrels of oil equivalent [BOE] (3.4×10^{10} GJ). The field was found in the deep waters off the northern coast of Egypt and ENI claims that it will be the largest ever in the Mediterranean and even the world.

Natural Gas as an Asset Class for Institutional Investors

Research conducted by the World Pensions Council (WPC) suggests that large US and Canadian pension funds and Asian and MENA area SWF investors have become particularly active in the fields of natural gas and natural gas infrastructure, a trend started in 2005 by the formation of Scotia Gas Networks in the UK by OMERS and Ontario Teachers' Pension Plan.

Adsorbed Natural Gas (ANG)

Natural gas may be stored by adsorbing it to the porous solids called sorbents. The optimal condition for methane storage is at room temperature and atmospheric pressure. Pressures up to 4 MPa (about 40 times atmospheric pressure) will yield greater storage capacity. The most common sorbent used for ANG is activated carbon (AC), primarily in three forms: Activated Carbon Fiber (ACF), Powdered Activated Carbon (PAC), activated carbon monolith.

RENEWABLE NATURAL GAS

Renewable Natural Gas (RNG), also known as Sustainable Natural Gas (SNG) or biomethane, is a biogas which has been upgraded to a quality similar to fossil natural gas and having a methane concentration of 90% or greater. A biogas is a gaseous form of methane obtained from biomass. By upgrading the quality to that of natural gas, it becomes possible to distribute the gas to customers via the existing gas grid within existing appliances. Renewable natural gas is a subset of synthetic natural gas or substitute natural gas (SNG).

Renewable natural gas can be produced and distributed via the existing gas grid, making it an attractive means of supplying existing premises with renewable heat and renewable gas energy, while requiring no extra capital outlay of the customer. The existing gas network also allows distribution of gas energy over vast distances at a minimal cost in energy. Existing networks would allow biogas to be sourced from remote markets that are rich in low-cost biomass (Russia or Scandinavia for example). Renewable natural gas can also be converted into liquefied natural gas (LNG) for direct use as fuel in transport sector.

The UK National Grid believes that at least 15% of all gas consumed could be made from matter such as sewage, food waste such as food thrown away by supermarkets and restaurants and organic waste created by businesses such as breweries. In the United States, analysis conducted in 2011 by the Gas Technology Institute determined that renewable gas from waste biomass including agricultural waste has the potential to add up to 2.5 quadrillion Btu annually, being enough to meet the natural gas needs of 50% of American homes.

In combination with power-to-gas, whereby the carbon dioxide and carbon monoxide fraction of biogas are converted to methane using electrolyzed hydrogen, the renewable gas potential of raw biogas is approximately doubled. Many ways of methanising carbon dioxide/monoxide and hydrogen exist, including biomethanation, the sabatier process and a new electrochemical process pioneered in the United States currently undergoing trials.

Manufacturing

A biomass to SNG efficiency of 70% can be achieved. Costs are minimised by maximising production scale and by locating an anaerobic digestion plant next to transport links (e.g. a port or river) for the chosen source of biomass. The existing gas storage infrastructure would allow the plant to continue to manufacture gas at the full utilisation rate even during periods of weak demand, helping minimise manufacturing capital costs per unit of gas produced.

Renewable gas can be produced through three main processes:

- Anaerobic digestion of organic (normally moist) material.

- Thermal gasification of organic (normally dry) material.

- Production through the Sabatier reaction. In these cases, the gas from primary production has to be upgraded in a secondary step to produce gas that is suitable for injection into the gas grid.

Commercial Development

BioSNG

Göteborg Energi opened the first demonstration plant for large scale production of bio-SNG through gasification of forest residues in Gothenburg, Sweden within the GoBiGas project. The plant had the capacity to produce 20 megawatts-worth of bioSNG from about 30 MW-worth of biomass, aiming at a conversion efficiency of 65%. From December 2014 the bioSNG plant was fully operational and supplied gas to the Swedish natural gas grid, reaching the quality demands with a methane content of over 95%. The plant was permanently closed due to economic problems in April 2018. Göteborg Energi had invested 175 million euro in the plant and intensive attempts for a year to sell the plant to new investors had failed.

It can be noted that the plant was a technical success, and performed as intended. However, natural gas is at a very low price given market conditions globally. It is expected the plant is to re-emerge around 2030 when economic conditions may be more favourable, with the possibility of a higher carbon price.

SNG is of particular interest in countries with extensive natural gas distribution networks. Core advantages of SNG include compatibility with existing natural gas infrastructure, higher efficiency that Fisher-Tropsch fuels production and smaller-production scale than other second generation biofuel production systems. The Energy Research Centre of the Netherlands has conducted extensive research on large-scale SNG production from woody biomass, based on the importation of feedstocks from abroad.

Renewable natural gas plants based on wood can be categorised into two main categories, one being allothermal, which has the energy provided by a source outside of the gasifier. One example is the double-chambered fluidised bed gasifiers consisting of a separate combustion and gasification chambers. Autothermal systems generate the heat within the gasifier, but require the use of pure oxygen to avoid nitrogen dilution.

In the UK, NNFCC found that any UK bioSNG plant built by 2020 would be highly likely to use 'clean woody feedstocks' and that there are several regions with good availability of that source.

Upgraded Biogas

In the UK, using anaerobic digestion is growing as a means of producing renewable biogas, with nearly 90 biomethane injection sites built across the country. Ecotricity announced plans to supply green gas to UK consumers via the national grid. Centrica also announced that it would begin injecting gas, manufactured from sewage, into the gas grid. In Canada, FortisBC, a gas provider in British Columbia, injects renewably created natural gas into its existing gas distribution system.

Sustainable Synthetic Natural Gas

Sustainable SNG is produced by high temperature oxygen blown slagging co-gasification at 70 to 75 bar pressure of biomass or waste residue. The advantage of a wide range of feedstocks is that much larger quantities of renewable SNG can be produced compared with biogas, with fewer supply chain limitations. A wide range of fuels with an overall biogenic carbon content of 50 to 55% is technically and financially viable. Hydrogen is added to the fuel mix during the gasification process, and carbon dioxide is removed by capture from the purge gas 'slip stream' syngas clean-up and catalytic methanation stages.

Large scale sustainable SNG will enable the UK gas and electricity grids to be substantially de-carbonised in parallel at source, while maintaining the existing operational and economic relationship between the gas and electricity grids. Carbon capture and sequestration can be added at little additional cost, thereby progressively achieving deeper de-carbonisation of the existing gas and electricity grids at low cost and operational risk. Cost benefit studies indicate that large scale 50% biogenic carbon content sustainable SNG can be injected into the high pressure gas transmission grid at a cost of around 65p/therm. At this cost, it is possible to re-process fossil natural gas, used as an energy input into the gasification process, into 5 to 10 times greater quantity of sustainable SNG. Large scale sustainable SNG, combined with continuing natural gas production from UK continental shelf and unconventional gas, will potentially enable the cost of UK peak electricity to be de-coupled from international oil denominated 'take or pay' gas supply contracts.

Applications:

- Electricity generation

- Space heating

- Process heating

- Biomass with carbon capture and storage

- Transportation fuel

Environment

Biogas creates similar environmental pollutants as ordinary natural gas fuel, such as carbon monoxide, sulfur dioxide, nitrogen oxide, hydrogen sulfide and particulates. Any unburned gas that escapes contains methane, a long lived greenhouse gas. The key difference from fossil natural gas is that it is often considered partly or fully carbon neutral, since the carbon dioxide contained in the biomass is naturally renewed in each generation of plants, rather than being released from fossil stores and increasing atmospheric carbon dioxide.

BIOGAS

Biogas is a type of biofuel that is naturally produced from the decomposition of organic waste. When organic matter, such as food scraps and animal waste, break down in an anaerobic environment (an environment absent of oxygen) they release a blend of gases, primarily methane and carbon dioxide. Because this decomposition happens in an anaerobic environment, the process of producing biogas is also known as anaerobic digestion.

Anaerobic digestion is a natural form of waste-to-energy that uses the process of fermentation to breakdown organic matter. Animal manure, food scraps, wastewater, and sewage are all examples of organic matter that can produce biogas by anaerobic digestion. Due to the high content of methane in biogas (typically 50-75%) biogas is flammable, and therefore produces a deep blue flame, and can be used as an energy source.

The Ecology of Biogas

Biogas is known as an environmentally-friendly energy source because it alleviates two major environmental problems simultaneously:

1. The global waste epidemic that releases dangerous levels of methane gas every day.

2. The reliance on fossil fuel energy to meet global energy demand.

By converting organic waste into energy, biogas is utilizing nature's elegant tendency to recycle substances into productive resources. Biogas generation recovers waste materials that would otherwise pollute landfills; prevents the use of toxic chemicals in sewage treatment plants, and saves money, energy, and material by treating waste on-site. Moreover, biogas usage does not require fossil fuel extraction to produce energy.

Instead, biogas takes a problematic gas, and converts it into a much safer form. More specifically, the methane content present in decomposing waste is converted into carbon dioxide. Methane gas has approximately 20 to 30 times the heat-trapping capabilities of carbon dioxide. This means that when a rotting loaf of bread converts into biogas, the loaf's environmental impact will be about 10 times less potent than if it was left to rot in a landfill.

Biogas Digesters

As opposed to letting methane gas release to the atmosphere, biogas digesters are the systems that process waste into biogas, and then channel that biogas so that the energy can be productively used. There are several types of biogas systems and plants that have been designed to make efficient use of biogas. While each model differs depending on input, output, size, and type, the biological process that converts organic waste into biogas is uniform. Biogas digesters receive organic matter, which decompose in a digestion chamber. The digestion chamber is fully submerged in water, making it an anaerobic (oxygen-free) environment. The anaerobic environment allows for microorganisms to break down the organic material, and convert it into biogas.

All-natural Fertilizer

Because the organic material decomposes in a liquid environment, nutrients present in the waste dissolve into the water, and create a nutrient-rich sludge, typically used as fertilizer for plants. This fertilizer output is generated on a daily basis, and therefore is a highly productive by-product of anaerobic digestion.

Biological Breakdown

To produce biogas, organic matter ferments with the help of bacterial communities. Four stages of fermentation move the organic material from their initial composition into their biogas state.

1. The first stage of the digestion process is the hydrolysis stage. In the hydrolysis stage insoluble organic polymers (such as carbohydrates) are broken down, making it accessible to the next stage of bacteria called acidogenic bacteria.

2. The acideogenic bacteria convert sugars and amino acids into carbon dioxide, hydrogen, ammonia, and organic acids.

3. At the third stage the acetogenic bacteria convert the organic acids into acetic acid, hydrogen, ammonia, and carbon dioxide, allowing for the final stage- the methanogens.

4. The methanogens convert these final components into methane and carbon dioxide- which can then be used as a flammable, green energy.

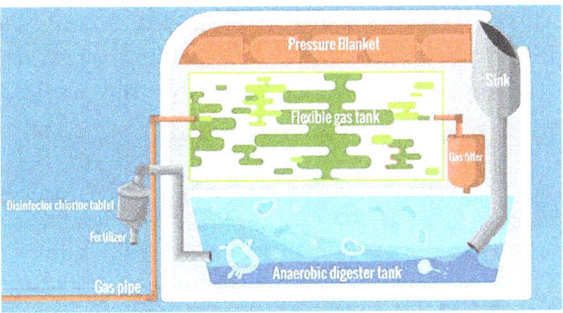

Many uses of Biogas

Biogas can be produced with various types of organic matter, and therefore there are several types of models for biogas digesters. Some industrial systems are designed to treat: municipal wastewater, industrial wastewater, municipal solid waste, and agricultural waste.

Small-scale systems are typically used for digesting animal waste. And newer family-size systems are designed to digest food waste. The resulting biogas can be used in several ways including: gas, electricity, heat, and transportation fuels.

For example, in Sweden hundreds of cars and buses run on refined biogas. The biogas in Sweden is produced primarily from sewage treatment plants and landfills.

Another example of the diversified uses of biogas is the First Milk plant. One of the UK's biggest cheese makers is building an anaerobic digestion plant that will process dairy residues and convert into bio-methane for the gas grid. New anaerobic digestion plants like these with fascinating stories keep popping up every day.

Small-scale Biogas Systems

Small-scale, or family-size biogas digesters are most frequently found in India and China. However, the demand for such units is growing rapidly throughout the world thanks to more advanced and convenient technologies, such as HomeBiogas. As the modern world is producing more and more waste, individuals are eager to find ecologic ways to treat their trash.

Traditional systems typically found in India and China focus on animal waste. Due to a lack of energy in rural areas combined with a surplus of animal manure, biogas digesters are very popular, useful, and even life-changing. In many developing countries, biogas digesters are even subsidized and advocated by the government and local ministries, who see the variety of benefits produced from using biogas. In addition to having a clean renewable energy provide gas in the kitchen, many families make extensive use of the fertilizer by-product that biogas digesters provide.

In African countries, some biogas users even turn a profit by selling the bio-slurry by-product produced by biogas systems. This bio-slurry is different from the liquid fertilizer that is produced daily. Bio-slurry refers to the most decomposed stage of the organic matter, after it has been broken down in the system. Bio-slurry sinks to the bottom of the biogas system, and with the help of modern units like HomeBiogas, is easily emptied out once accrued (usually an annual process). This bio-slurry is in fact a nutrient-dense sludge that provides lots of benefits to soil, and can increase productivity of vegetable gardens.

Biogas is a technology that mimics nature's ability to give back. Both industrial-size and family-size biogas units are becoming incredibly popular and relevant in today's world. As the application and efficiency grows, biogas can make a significant impact on reducing greenhouse gases. As a clean source of energy and a renewable means of treating organic waste, biogas is applicable both in under-developed and industrialized countries.

BIOCHAR

Biochar is charcoal used as a soil amendment. Biochar is a stable solid, rich in carbon, and can endure in soil for thousands of years. Like most charcoal, biochar is made from biomass via pyrolysis. Biochar is under investigation as an approach to carbon sequestration, as it has the potential to help mitigate climate change. It results from processes related to pyrogenic carbon capture and storage (PyCCS). Independently, biochar can increase soil fertility of acidic soils (low pH soils), increase agricultural productivity, and provide protection against some foliar and soil-borne diseases. Regarding the definition from the production part, biochar is defined by the International Biochar Initiative as "The solid material obtained from the thermochemical conversion of biomass in an oxygen-limited environment".

Production

Biochar is a high-carbon, fine-grained residue that today is produced through modern pyrolysis processes; it is the direct thermal decomposition of biomass in the absence of oxygen (preventing combustion), which produces a mixture of solids (the biochar proper), liquid (bio-oil), and gas (syngas) products. The specific yield from the pyrolysis is dependent on process condition, such as temperature, residence time and heating rate. These parameters can be optimized to produce either energy or biochar. Temperatures of 400–500 °C (673–773 K) produce more char, while temperatures above 700 °C (973 K) favor the yield of liquid and gas fuel components. Pyrolysis occurs more quickly at the higher temperatures, typically requiring seconds instead of hours. The increasing heating rate will also lead to a decrease of pyrolysis biochar yield, while the temperature is in the range of 350–600 °C (623–873 K). Typical yields are 60% bio-oil, 20% biochar, and 20% syngas. By comparison, slow pyrolysis can produce substantially more char (≈35%); it is this which contributes to the observed soil fertility of terra preta. Once initialized, both processes produce net energy. For typical inputs, the energy required to run a "fast" pyrolyzer is approximately 15% of the energy that it outputs. Modern pyrolysis plants can use the syngas created by the pyrolysis process and output 3–9 times the amount of energy required to run.

Besides pyrolysis, torrefaction and hydrothermal carbonization process can also thermally decompose biomass to the solid material. However, these products cannot be strictly defined as biochar. The carbon product from the torrefaction process still remains some volatile organic components, thus its properties are between that of biomass feedstock and biochar. Furthermore, even the hydrothermal carbonization could produce a carbon-rich solid product, the hydrothermal carbonization is evidently different from the conventional thermal conversion process. Therefore, the solid product from hydrothermal carbonization is defined as "hydrochar" rather than "biochar".

The Amazonian pit/trench method harvests neither bio-oil nor syngas, and releases a large amount of CO_2, black carbon, and other greenhouse gases (GHGs) (and potentially, toxins) into the air, though less greenhouse gasses than captured during the growth of the biomass. Commercial-scale systems process agricultural waste, paper byproducts, and even municipal waste and typically eliminate these side effects by capturing and using the liquid and gas products. The production of biochar as an output is not a priority in most cases.

Centralized, Decentralized and Mobile Systems

In a centralized system, all biomass in a region is brought to a central plant (i.e. biomass-fueled thermal power station) for processing into biochar. Alternatively, each farmer or group of farmers can operate a lower-tech kiln. Finally, a truck equipped with a pyrolyzer can move from place to place to pyrolyze biomass. Vehicle power comes from the syngas stream, while the biochar remains on the farm. The biofuel is sent to a refinery or storage site. Factors that influence the choice of system type include the cost of transportation of the liquid and solid byproducts, the amount of material to be processed, and the ability to feed directly into the power grid.

The most common crops used for making biochar include various tree species, as well as various energy crops. Some of these energy crops (i.e. Napier grass) can also store much more carbon on a shorter timespan than trees do.

For crops that are not exclusively for biochar production, the Residue-to-Product Ratio (RPR) and the collection factor (CF) the percent of the residue not used for other things, measure the approximate amount of feedstock that can be obtained for pyrolysis after harvesting the primary product. For instance, Brazil harvests approximately 460 million tons (MT) of sugarcane annually, with an RPR of 0.30, and a CF of 0.70 for the sugarcane tops, which normally are burned in the field. This translates into approximately 100 MT of residue annually, which could be pyrolyzed to create energy and soil additives. Adding in the bagasse (sugarcane waste) (RPR=0.29 CF=1.0), which is otherwise burned (inefficiently) in boilers, raises the total to 230 MT of pyrolysis feedstock. Some plant residue, however, must remain on the soil to avoid increased costs and emissions from nitrogen fertilizers.

Pyrolysis technologies for processing loose and leafy biomass produce both biochar and syngas.

Thermo-catalytic Depolymerization

Alternatively, "thermo-catalytic depolymerization", which utilizes microwaves, has recently been used to efficiently convert organic matter to biochar on an industrial scale, producing $\approx$50% char.

Properties

The physical and chemical properties of biochars as determined by feedstocks and technologies are crucial for the application of biochars in the industry and environment. Different characterization data are employed to biochars and determine their performances in a specific use. For example, the guidelines published by the International Biochar Initiative provide standardized methods in evaluating the product quality of biochar for soil application. The properties of biochar can be characterized in several respects, including the proximate and elemental composition, pH value, porosity etc., which correlate with different biochar properties. The atomic ratios of biochar, including H/C and O/C, correlate with the biochar properties that are relevant to the organic content such as polarity and aromaticity. The van-Krevelen diagram can be used to show the evolution of biochar atomic ratios in the production process. In the carbonization process, both the H/C and O/C ratio decreased due to the release of functional groups which contain hydrogen and oxygen.

Uses

Carbon Sink

The burning and natural decomposition of biomass and in particular agricultural waste adds large amounts of CO_2 to the atmosphere. Biochar is a stable way of storing carbon in the ground for centuries, potentially reducing or stalling the growth in atmospheric greenhouse gas levels; at the same time its presence in the earth can improve water quality, increase soil fertility, raise agricultural productivity, and reduce pressure on old-growth forests.

Biochar can sequester carbon in the soil for hundreds to thousands of years, like coal. Such a carbon-negative technology would lead to a net withdrawal of CO_2 from the atmosphere, while producing consumable energy. This technique is advocated by prominent scientists such as James Hansen, head of the NASA Goddard Institute for Space Studies, and James Lovelock, creator of the Gaia hypothesis, for mitigation of global warming by greenhouse gas remediation.

Researchers have estimated that sustainable use of biocharring could reduce the global net emissions of carbon dioxide (CO_2), methane, and nitrous oxide by up to 1.8 Pg CO_2-C equivalent (CO_2-C_e) per year (12% of current anthropogenic CO_2-C_e emissions; 1 Pg=1 Gt), and total net emissions over the course of the next century by $130\,\mathrm{Pg}\ CO_2\text{-}C_e$, without endangering food security, habitat, or soil conservation.

Soil Amendment

Biochar is recognized as offering a number of benefits for soil health. Many benefits are related to the extremely porous nature of biochar. This structure is found to be very effective at retaining both water and water-soluble nutrients. Soil biologist Elaine Ingham indicates the extreme suitability of biochar as a habitat for many beneficial soil micro organisms. She points out that when pre-charged with these beneficial organisms biochar becomes an extremely effective soil amendment promoting good soil and, in turn, plant health.

Biochar has also been shown to reduce leaching of *E-coli* through sandy soils depending on application rate, feedstock, pyrolysis temperature, soil moisture content, soil texture, and surface properties of the bacteria.

For plants that require high potash and elevated pH, biochar can be used as a soil amendment to improve yield.

Biochar can improve water quality, reduce soil emissions of greenhouse gases, reduce nutrient leaching, reduce soil acidity, and reduce irrigation and fertilizer requirements. Biochar was also found under certain circumstances to induce plant systemic responses to foliar fungal diseases and to improve plant responses to diseases caused by soilborne pathogens.

The various impacts of biochar can be dependent on the properties of the biochar, as well as the amount applied, and there is still a lack of knowledge about the important mechanisms and properties. Biochar impact may depend on regional conditions including soil type, soil condition (depleted or healthy), temperature, and humidity. Modest additions of biochar to soil reduce nitrous oxide N_2O emissions by up to 80% and eliminate methane emissions, which are both more potent greenhouse gases than CO_2.

Studies have reported positive effects from biochar on crop production in degraded and nutrient–poor soils. The application of compost and biochar under FP7 project FERTIPLUS has had positive effects in soil humidity, and crop productivity and quality in different countries. Biochar can be designed with specific qualities to target distinct properties of soils. In a Colombian savanna soil, biochar reduced leaching of critical nutrients, created a higher crop uptake of nutrients, and provided greater soil availability of nutrients. At 10% levels biochar reduced contaminant levels in plants by up to 80%, while reducing total chlordane and DDX content in the plants by 68 and 79%, respectively. On the other hand, because of its high adsorption capacity, biochar may reduce the efficacy of soil applied pesticides that are needed for weed and pest control. High-surface-area biochars may be particularly problematic in this regard; more research into the long-term effects of biochar addition to soil is needed.

Slash-and-char

Switching from *slash-and-burn* to *slash-and-char* farming techniques in Brazil can decrease both deforestation of the Amazon basin and carbon dioxide emission, as well as increase crop yields. Slash-and-burn leaves only 3% of the carbon from the organic material in the soil.

Slash-and-char can keep up to 50% of the carbon in a highly stable form. Returning the biochar into the soil rather than removing it all for energy production reduces the need for nitrogen fertilizers, thereby reducing cost and emissions from fertilizer production and transport. Additionally, by improving the soil's ability to be tilled, its fertility and its productivity, biochar-enhanced soils can indefinitely sustain agricultural production, whereas non-enriched soils quickly become depleted of nutrients, forcing farmers to abandon the fields, producing a continuous slash and burn cycle and the continued loss of tropical rainforest. Using pyrolysis to produce bio-energy also has the added benefit of not requiring infrastructure changes the way processing biomass for cellulosic ethanol does. Additionally, the biochar produced can be applied by the currently used machinery for tilling the soil or equipment used to apply fertilizer.

Water Retention

Biochar is hygroscopic. Thus it is a desirable soil material in many locations due to its ability to attract and retain water. This is possible because of its porous structure and high specific surface area. As a result, nutrients such as phosphate, and agrochemicals are retained for the plants benefit. Plants are therefore healthier, and less fertilizer leaches into surface or groundwater.

Energy Production: Bio-oil and Syngas

Mobile pyrolysis units can be used to lower the costs of transportation of the biomass if the biochar is returned to the soil and the syngas stream is used to power the process. Bio-oil contains organic acids that are corrosive to steel containers, has a high water vapor content that is detrimental to ignition, and, unless carefully cleaned, contains some biochar particles which can block injectors. Currently, it is less suitable for use as a kind of biodiesel than other sources.

If biochar is used for the production of energy rather than as a soil amendment, it can be directly substituted for any application that uses coal. Pyrolysis also may be the most cost-effective way of electricity generation from biomaterial.

Direct and Indirect Benefits

- The pyrolysis of forest- or agriculture-derived biomass residue generates a biofuel without competition with crop production.

- Biochar is a pyrolysis byproduct that may be ploughed into soils in crop fields to enhance their fertility and stability, and for medium- to long-term carbon sequestration in these soils. It has meant a remarkable improvement in tropical soils showing positive effects in increasing soil fertility and in improving disease resistance in West European Soils.

- Biochar enhances the natural process: the biosphere captures CO_2, especially through plant production, but only a small portion is stably sequestered for a relatively long time (soil, wood, etc.).

- Biomass production to obtain biofuels and biochar for carbon sequestration in the soil is a carbon-negative process, i.e. more CO_2 is removed from the atmosphere than released, thus enabling long-term sequestration.

Research

Intensive research into manifold aspects involving the pyrolysis/biochar platform is underway around the world. From 2005 to 2012, there were 1,038 articles that included the word "biochar" or "bio-char" in the topic that had been indexed in the ISI Web of Science. Further research is in progress by such diverse institutions around the world as Cornell University, the University of Edinburgh (which has a dedicated research unit), University of Georgia the Agricultural Research Organization (ARO) of Israel, Volcani Center, where a network of researchers involved in biochar research (iBRN, Israel Biochar Researchers Network) was established as early as 2009, and the University of Delaware.

Long-term effect of biochar on soil C sequestration of recent carbon inputs has been examined using soil from arable fields in Belgium with charcoal-enriched black spots dating >150 years ago from historical charcoal production mound kilns. Topsoils from these 'black spots' had a higher organic C concentration [3.6 ± 0.9% organic carbon (OC)] than adjacent soils outside these black spots (2.1 ± 0.2% OC). The soils had been cropped with maize for at least 12 years which provided a continuous input of C with a C isotope signature ($\delta13C$) −13.1, distinct from the $\delta13C$ of soil organic carbon (−27.4 ‰) and charcoal (−25.7 ‰) collected in the surrounding area. The isotope signatures in the soil revealed that maize-derived C concentration was significantly higher in charcoal-amended samples ('black spots') than in adjacent unamended ones (0.44% vs. 0.31%; $P = 0.02$). Topsoils were subsequently collected as a gradient across two 'black spots' along with corresponding adjacent soils outside these black spots and soil respiration, and physical soil fractionation was conducted. Total soil respiration (130 days) was unaffected by charcoal, but the maize-derived C respiration per unit maize-derived OC in soil significantly decreased about half ($P < 0.02$) with increasing charcoal-derived C in soil. Maize-derived C was proportionally present more in protected soil aggregates in the presence of charcoal. The lower specific mineralization and increased C sequestration of recent C with charcoal are attributed to a combination of physical protection, C saturation of microbial communities and, potentially, slightly higher annual primary production. Overall, this study provides evidence of the capacity of biochar to enhance C sequestration in soils through reduced C turnover on the long term.

Biochar sequesters carbon (C) in soils because of its prolonged residence time, ranging from several years to millennia. In addition, biochar can promote indirect C-sequestration by increasing crop yield while, potentially, reducing C-mineralization. Laboratory studies have evidenced effects of biochar on C-mineralization using 13C isotope signatures.

Fluorescence analysis of the dissolved organic matter from soil amended with biochar revealed that biochar application increased a humic-like fluorescent component, likely associated with biochar-carbon in solution. The combined spectroscopy-microscopy approach revealed the accumulation of aromatic-carbon in discrete spots in the solid-phase of microaggregates and its co-localization with clay minerals for soil amended with raw residue or biochar. The co-localization of aromatic-C:polysaccharides-C was consistently reduced upon biochar application. These finding suggested that reduced C metabolism is an important mechanism for C stabilization in biochar-amended soils.

Students at Stevens Institute of Technology in New Jersey are developing supercapacitors that use electrodes made of biochar. A process developed by University of Florida researchers that removes phosphate from water, also yields methane gas usable as fuel and phosphate-laden carbon suitable for enriching soil. Researchers at the University of Auckland are also working on utilizing biochar in concrete applications to reduce carbon emissions during concrete production and to improve the strength considerably. It has also demonstrated that the biochar can be used as a suitable filler in polymer matrix. Recently, biochar-starch bio-composites were prepared and its nano-mechanical behavious were investigated using advanced dynamic atomic force microscopy.

BIOGASOLINE

Biogasoline is gasoline produced from biomass such as algae. Like traditionally produced gasoline, it contains between 6 (hexane) and 12 (dodecane) carbon atoms per molecule and can be used in internal-combustion engines. Biogasoline is chemically different from biobutanol and bioethanol, as these are alcohols, not hydrocarbons.

Companies such as Diversified Energy Corporation are developing approaches to take triglyceride inputs and through a process of deoxygenation and reforming (cracking, isomerizing, aromatizing, and producing cyclic molecules) producing biogasoline. This biogasoline is intended to match the chemical, kinetic, and combustion characteristics of its petroleum counterpart, but with much higher octane levels. Others are pursuing similar approaches based on hydrotreating. And lastly still others are focused on the use of woody biomass for conversion to biogasoline using enzymatic processes.

Structure and Properties

BG100, or 100% biogasoline, can immediately be used as a drop-in substitute for petroleum gasoline in any conventional gasoline engine, and can be distributed in the same fueling infrastructure, as the properties match traditional gasoline from petroleum. Dodecane requires a small percentage of octane booster to match gasoline. Ethanol fuel (E85) requires a special engine and has lower combustion energy and corresponding fuel economy.

But due to biogasoline's chemical similarities it can also be mixed with regular gasoline. You can have higher ratios of biogasoline to gasoline and not have to modify the vehicles engine unlike ethanol.

Comparison to Common Fuels

Fuel	Energy Density MJ/L	Air-fuel ratio	Specific Energy MJ/kg	Heat of Vaporization MJ/kg	RON	MON
Gasoline	34.6	14.6	46.9	0.36	91–99	81–89
Butanol fuel	29.2	11.2	36.6	0.43	96	78
Ethanol fuel	24.0	9.0	30.0	0.92	129	102
Methanol fuel	19.7	6.5	15.6	1.2	136	104

Production

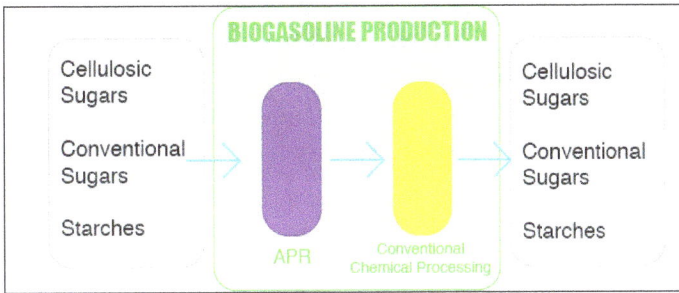

Biogasoline Production Process.

Biogasoline is created by turning sugar directly into gasoline. In late March 2010, the world's first biogasoline demonstration plant was started in Madison, WI by Virent Energy Systems, Inc. Virent discovered and developed a technique called Aqueous Phase Reforming (APR) in 2001. APR includes many processes including reforming to generate hydrogen, dehydrogenation of alcohols/hydrogenation of carbonyls, deoxygenation reactions, hydrogenolysis and cyclization. The input for APR is a carbohydrate solution created from plant material, and the product is a mixture of chemicals and oxygenated hydrocarbons. From there, the materials go through further conventional chemical processing to yield the final result: a mixture of non-oxygenated hydrocarbons that they claimed was cost-effective. These hydrocarbons are the exact hydrocarbons found in petroleum fuels which is why today's cars do not need to be altered to run on biogasoline. The only difference is in origin. Petroleum based fuels are made from oil, and biogasoline is made from plants such as beets and sugarcane or cellulosic biomass which would normally be plant waste.

Diesel fuel is made up of linear hydrocarbons. These are long straight carbon atom chains. They differ from the shorter, branched hydrocarbons that make up gasoline. In 2014 Researchers used a feedstock of levulinic acid to create biogasoline. Levulinic acid is derived from cellulose material, such as corn stalks, straw or other plant waste. That waste does not have to be fermented. The fuel-making process is reportedly inexpensive and offers yields of over 60 percent.

Economic Viability and Future

One of the major problems facing the economic viability of biogasoline is the high up- front cost. Research groups are finding that current investment groups are impatient with the pace of biogasoline progress. In addition, environmental groups may demand that biogasoline that is produced in a way that protects wildlife, especially fish. A research group studying the economic viability of biofuels found that current techniques of production and high costs of production will prevent biogasoline from being accessible to the general public. The group determined that the price of biogasoline would need to be approximately $800 per barrel, which they determine as unlikely with current production costs. Another problem inhibiting the success of biogasoline is the lack of tax relief. The government is providing tax relief for ethanol fuels but has yet to offer tax relief for biogasoline. This makes biogasoline a much less attractive option to consumers. Lastly, producing biogasoline could have a large effect on the farming industry. If biogasoline became a serious alternative, a large percentage of our existing arable land would be converted to grow crops solely for biogasoline. This could decrease the amount of land used to farm food for human consumption and may decrease overall feedstock. This would cause an increase in overall food cost.

While there may be some problems facing the economic viability of biogasoline, the partnership between Royal Dutch Shell and Virent Energy Systems, Inc., a bioscience firm based in Madison, WI, to further research biogasoline is an encouraging sign for biogasoline's future. In addition, many nations are enacting policies that increase the use of biogasoline within the country to help curb the cost of fossil fuels and create more energy independence. Current efforts by the partnership are focused on improving the technology and making it available for large-scale production.

BIOLIQUIDS

Bioliquids are liquid fuels made from biomass for energy purposes other than transport (i.e. heating and electricity).

Bioliquids are usually made from virgin or used vegetable and seed oils, like palm or soya oil. These oils are burned in a power station to create heat, which can then be used to warm homes or boil water to make steam. This steam can then be used to drive a turbine to generate electricity.

Rudolf Diesel's first public exhibition of the internal combustion engine, that was to later bear his name, ran on peanut oil.

Bioliquid Production and use

Bioliquids have been used for many years to provide heat for homes on a small scale but now big energy providers are looking at their use on a larger scale.

A controversial plant in Bristol (UK) was recently given the go ahead despite receiving several hundred complaints. The plant will be built and operated by W4B and provide enough power for 25,000 homes.

Advantages

Bioliquids have several key advantages over other sources of renewable energy:

- Bioliquids have a high energy density.

- The technology is well established, having been used for many years.

- Can be used on demand, reacting quickly to changes in demand for power.

- Can help reduce dependency on foreign oil.

- Reduces the green house gas emissions.

Disadvantages

Many of the same problems that affect biofuels also affect bioliquids and there are various social, economic, environmental and technical issues, These include: the effect of moderating oil prices, the "food vs fuel" debate, poverty reduction potential, carbon emissions levels, sustainable biofuel production, deforestation and soil erosion, loss of biodiversity, impact on water resources, as well as energy balance and efficiency.

Bioliquids also have several key problems compared to other sources of renewable energy:

- Price of fuel is very variable, due to competitiveness of feedstock for other uses (e.g. soap).

- Supply chain is still very new.

- Governments, such as the EU, remained undecided on bioliquids.

BAGASSE

Bagasse is the dry pulpy fibrous residue that remains after sugarcane or sorghum stalks are crushed to extract their juice. It is used as a biofuel for the production of heat, energy, and electricity, and in the manufacture of pulp and building materials.

Agave bagasse is a similar material that consists of the tissue of the blue agave after extraction of the sap.

Production, Storage and Composition

Sugarcane being crushed in Engenho da Calheta, Madeira. The bagasse falls down a chute and is removed on a conveyor belt below.

For every 10 tonnes of sugarcane crushed, a sugar factory produces nearly three tonnes of wet bagasse. Since bagasse is a by-product of the cane sugar industry, the quantity of production in each country is in line with the quantity of sugarcane produced.

The high moisture content of bagasse, typically 40–50 percent, is detrimental to its use as a fuel. In general, bagasse is stored prior to further processing. For electricity production, it is stored under moist conditions, and the mild exothermic process that results from the degradation of residual sugars dries the bagasse pile slightly. For paper and pulp production, it is normally stored wet in order to assist in removal of the short pith fibres, which impede the paper making process, as well as to remove any remaining sugar.

A typical chemical analysis of washed and dried bagasse might show:

- Cellulose 45–55 percent.

- Hemicellulose 20–25 percent.

- Lignin 18–24 percent.

- Ash 1–4 percent.

- Waxes <1 percent.

Bagasse is a heterogeneous material containing around 30-40 percent of "pith" fibre, which is derived from the core of the plant and is mainly parenchyma material, and "bast", "rind", or "stem" fibre, which makes up the balance and is largely derived from sclerenchyma material. These properties make bagasse particularly problematic for paper manufacture and have been the subject of a large body of literature.

Uses

Bagasse, covered with blue plastic, outside a sugar mill in Proserpine.

Many research efforts have explored using bagasse as a biofuel in renewable power generation and in the production of bio-based materials.

Fuel

Bagasse is often used as a primary fuel source for sugar mills. When burned in quantity, it produces sufficient heat energy to supply all the needs of a typical sugar mill, with energy to spare. To this end, a secondary use for this waste product is in cogeneration, the use of a fuel source to provide both heat energy, used in the mill, and electricity, which is typically sold on to the consumer electrical grid.

The lower calorific value (LCV) of bagasse in kJ/kg may be estimated using the formula: $LCV = 18260$, where the moisture, brix and ash content of the bagasse are expressed as a percentage by mass. Similarly, the higher calorific value (HCV) of bagasse may be estimated using: $HCV = 19605 - 196.05 \times Moisture - 31.14 \times Brix - 196.05 \times Ash$.

The resulting CO_2 emissions are less than the amount of CO_2 that the sugarcane plant absorbed from the atmosphere during its growing phase, which makes the process of cogeneration greenhouse-gas-neutral. In countries such as Australia, sugar factories contribute "green" power to the electricity grid. Hawaiian Electric Industries also burns bagasse for cogeneration.

Ethanol produced from the sugar in sugarcane is a popular fuel in Brazil. The cellulose-rich bagasse is being widely investigated for its potential for producing commercial quantities of cellulosic ethanol. For example, until May 2015 BP was operating a cellulosic ethanol demonstration plant based on cellulosic materials in Jennings, Louisiana.

Bagasse's potential for advanced biofuels has been shown by several researchers. However, the compatibility with conventional fuels and suitability of these crude fuels in conventional engines have yet to be proven.

Pulp, Paper, Board and Feed

Bagasse is commonly used as a substitute for wood in many tropical and subtropical countries for the production of pulp, paper and board, such as India, China, Colombia, Iran, Thailand, and Argentina. It produces pulp with physical properties that are well suited for generic printing and writing papers as well as tissue products but it is also widely used for boxes and newspaper production. It can also be used for making boards resembling plywood or particle board, called bagasse board and Xanita board, and is considered a good substitute for plywood. It has wide usage for making partitions and furniture.

The industrial process to convert bagasse into paper was developed in 1937 in a small laboratory in Hacienda Paramonga, a sugar mill on the coast of Peru owned by W.R. Grace Company. With a promising method, the company bought an old paper mill in Whippany, New Jersey and shipped bagasse from Peru to test the viability of the process on an industrial scale. The first paper manufacturing machines were designed in Germany and installed in the Cartavio sugar cane plant in 1938. Sociedad Paramonga was bought in 1997 by Quimpac and in 2015 produced 90,000 metric tons of office paper, toilet paper and cardboard for the Peruvian market.

K-Much Industry has patented a method of converting bagasse into cattle feed by mixing it with molasses and enzymes (such as bromelain) and fermenting it. It is marketed in Thailand, Japan, Malaysia, Korea, Taiwan and Middle East and Australia.

Xanita, a South African company, mixes 30 percent bagasse cellulose fibres in with recycled kraft paper fibre to make ultra-lightweight composite boards. These are sold as an environmentally friendly, formaldehyde-free alternative to MDF and particle board.

Nanocellulose

Nanocellulose can be produced from bagasse through various conventional and novel processes. This provides a pathway to generate higher-value products from what can be considered a process waste stream.

Health Impact

Workplace exposure to dust from the processing of bagasse can cause the chronic lung condition pulmonary fibrosis, more specifically referred to as bagassosis.

Human Consumption

Processed bagasse is added to human food as sugarcane fiber. It is a soluble fiber but can help promote intestinal regularity. One animal study suggests that sugarcane fiber combined with a high-fat diet may help control type 2 diabetes. Bagasse are good sources of lignoceric and cerotic acids.

In Guangxi Zhuang Autonomous Region, China, bagasse is sometimes used to smoke bacon and sausages.

WOOD GAS

Wood gas is a syngas fuel which can be used as a fuel for furnaces, stoves and vehicles in place of gasoline, diesel or other fuels. During the production process biomass or other carbon-containing materials are gasified within the oxygen-limited environment of a wood gas generator to produce hydrogen and carbon monoxide. These gases can then be burnt as a fuel within an oxygen rich environment to produce carbon dioxide, water and heat. In some gasifiers this process is preceded by pyrolysis, where the biomass or coal is first converted to char, releasing methane and tar rich in polycyclic aromatic hydrocarbons.

Usage

Internal Combustion Engine

A Wood gas generator fitted to a Ford truck converted, into a tractor,
Per Larsen Tractor Museum.

Wood gasifiers can power either spark ignition engines, where all of the normal fuel can be replaced with little change to the carburation, or in a Diesel engine, feeding the gas into the air inlet that is modified to have a throttle valve, if it didn't have it already. On Diesel engines the Diesel fuel is still needed to ignite the gas mixture, so a mechanically regulated Diesel engine's "stop" linkage and probably "throttle" linkage must be modified to always give the engine a little bit of injected fuel, often under the standard idle per-injection volume. Wood can be used to power cars with ordinary internal combustion engines if a wood gasifier is attached. This was quite popular during World War II in several European, African and Asian countries, because the war prevented easy and cost-effective access to oil. In more recent times, wood gas has been suggested as a clean and efficient method to heat and cook in developing countries, or even to produce electricity when combined with an internal combustion engine. Compared to World War II technology, gasifiers have become less dependent on constant attention due to the use of sophisticated electronic control systems, but it remains difficult to get clean gas from them. Purification of the gas and feeding it into natural gas pipelines is one variant to link it to the existing refueling infrastructure. Liquefaction by the Fischer–Tropsch process is another possibility.

Wood gasifier system.

Efficiency of the gasifier system is relatively high. The gasification stage converts about 75% of fuel energy content into a combustible gas that can be used as a fuel for internal combustion engines. Based on long-term practical experiments and over 100,000 kilometres (62,000 mi) driven with a wood gas-powered car, the energy consumption has been 1.54 times higher compared to the energy demand of the same car on petrol, excluding the energy needed to extract, transport and refine the oil from which petrol is derived, and excluding the energy to harvest, process, and transport the wood to feed the gasifier. This means that 1,000 kilograms (2,200 lb) of wood combustible matter has been found to be equivalent to 365 litres (96 US gal) of petrol during real transportation in similar driving conditions and with the same, otherwise unmodified, vehicle. This can be considered to be a good result, because no other refining of the fuel is required. This study also considers all possible losses of the wood gas system, like preheating of the system and carrying of the extra weight of the gas-generating system. In power generation, reported demand of fuel is 1.1 kilograms (2.4 lb) wood combustible matter per kilowatt-hour of electricity.

A wood-gas powered car, Berlin, 1946. Note the secondary radiator, required to cool the gas before it's introduced into the engine.

Gasifiers have been built for remote Asian communities using rice hulls, which in many cases have no other use. One installation in Burma uses an 80 kW modified Diesel-powered electric generator for about 500 people who are otherwise without power. The ash can be used as biochar fertilizer, so this can be considered a renewable fuel.

Exhaust gas emission from an internal combustion engine is significantly lower on wood gas than on petrol. Especially the hydrocarbon emissions are low on wood gas. A normal catalytic converter works well with wood gas, but even without it, emission levels less than 20 ppm HC and 0.2% CO can be easily achieved by most automobile engines. Combustion of wood gas generates no particulates, and the gas renders thus very little carbon black amongst motor oil.

Stoves, Cooking and Furnaces

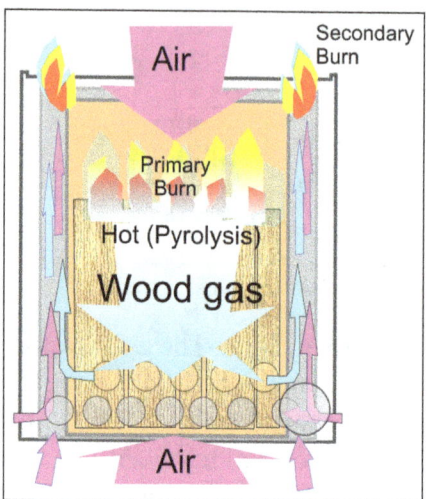

Coaxial downdraft gasification stove.

Certain stove designs are, in effect, gasifiers working on the updraft principle: The air passes up through the fuel, which can be a column of rice hulls, and is combusted, then reduced to carbon monoxide by the residual char on the surface. The resulting gas is then burnt by heated secondary air coming up a concentric tube. Such a device behaves very much like a gas stove. This arrangement is also known as a Chinese burner.

An alternative stove based on the down-draft principle and typically built with nested cylinders also provides high efficiency. Combustion from the top creates a gasification zone, with the gas escaping downwards through ports located at the base of the burner chamber. The gas mixes with additional incoming air to provide a secondary burn. Most of the CO produced by gasification is oxidized to CO_2 in the secondary combustion cycle; therefore, gasification stoves carry lower health risks than conventional cooking fires.

Another application is the use of producer gas to displace light density fuel oil (LDO) in industrial furnaces.

Production

A wood gasifier takes wood chips, sawdust, charcoal, coal, rubber or similar materials as fuel and burns these incompletely in a fire box, producing wood gas, solid ash and soot, the latter of which have to be removed periodically from the gasifier. The wood gas can then be filtered for tars and soot/ash particles, cooled and directed to an engine or fuel cell. Most of these engines have strict purity requirements of the wood gas, so the gas often has to pass through extensive gas cleaning in order to remove or convert, *i.e.*, "crack", tars and particles. The removal of tar is often accomplished by using a water scrubber. Running wood gas in an unmodified gasoline-burning internal combustion engine may lead to problematic accumulation of unburned compounds.

Fluidized bed gasifier in Güssing, Austria, operated on wood chips.

The quality of the gas from different gasifiers varies a great deal. Staged gasifiers, where pyrolysis and gasification occur separately, instead of in the same reaction zone as was the case in, *e.g.*, the World War II gasifiers, can be engineered to produce essentially tar-free gas (less than 1 mg/m³), while single-reactor fluid-bed gasifiers may exceed 50,000 mg/m³ tar. The fluid bed reactors have the advantage of being much more compact, with more capacity per unit volume and price. Depending on the intended use of the gas, tar can be beneficial, as well by increasing the heating value of the gas.

A charcoal gas producer at the Nambassa alternative festival in New Zealand.

During the production of charcoal for blackpowder, the volatile wood gas is vented. Extremely-high-surface-area carbon results, suitable for use as a fuel in black powder.

The heat of combustion of "producer gas" — a term used in the United States meaning wood gas produced for use in a combustion engine — is rather low compared to other fuels. Taylor reports that producer gas has a lower heat of combustion of 5.7 MJ/kg versus 55.9 MJ/kg for natural gas and 44.1 MJ/kg for gasoline. The heat of combustion of wood is typically 15-18 MJ/kg. Presumably, these values can vary somewhat from sample to sample. The same source reports the following chemical composition by volume which most likely is also variable:

- Nitrogen N_2: 50.9%

- Carbon monoxide CO: 27.0%

- Hydrogen H_2: 14.0%

- Carbon dioxide CO_2: 4.5%

- Methane CH_4: 3.0%

- Oxygen O_2: 0.6%

It is pointed out that the gas composition is strongly dependent on the gasification process, the gasification medium (air, oxygen or steam) and the fuel moisture. Steam-gasification processes typically yield high hydrogen contents, downdraft fixed bed gasifiers yield high nitrogen concentrations and low tar loads, while updraft fixed bed gasifiers yield high tar loads.

WOOD FUEL

Wood fuel (or fuelwood) is a fuel, such as firewood, charcoal, chips, sheets, pellets, and sawdust. The particular form used depends upon factors such as source, quantity, quality and application. In many areas, wood is the most easily available form of fuel, requiring no tools in the case of picking up dead wood, or few tools, although as in any industry, specialized tools, such as skidders and hydraulic wood splitters, have been developed to mechanize production. Sawmill waste and construction industry by-products also include various forms of lumber tailings. The discovery of how to make fire for the purpose of burning wood is regarded as one of humanity's most important advances. The use of wood as a fuel source for heating is much older than civilization and is assumed to have been used by Neanderthals. Today, burning of wood is the largest use of energy derived from a solid fuel biomass. Wood fuel can be used for cooking and heating, and occasionally for fueling steam engines and steam turbines that generate electricity. Wood may be used indoors in a furnace, stove, or fireplace, or outdoors in furnace, campfire, or bonfire.

Wood has been used as fuel for millennia. Historically, it was limited in use only by the distribution of technology required to make a spark. Heat derived from wood is still common throughout much of the world. Early examples included a fire constructed inside a tent. Fires were constructed on the ground, and a smoke hole in the top of the tent allowed the smoke to escape by convection.

In permanent structures and in caves, hearths were constructed or established—surfaces of stone or another noncombustible material upon which a fire could be built. Smoke escaped through a smoke hole in the roof.

In contrast to civilizations in relatively arid regions (such as Mesopotamia and Egypt), the Greeks, Romans, Celts, Britons, and Gauls all had access to forests suitable for using as fuel. Over the centuries there was a partial deforestation of climax forests and the evolution of the remainder to coppice with standards woodland as the primary source of wood fuel. These woodlands involved a continuous cycle of new stems harvested from old stumps, on rotations between seven and thirty years. One of the earliest printed books on woodland management, in English, was John Evelyn's "Sylva, or a discourse on forest trees" advising landowners on the proper management of forest estates. H. L. Edlin, in "Woodland Crafts in Britain", 1949 outlines the extraordinary techniques employed, and range of wood products that have been produced from these managed forests since pre-Roman times. And throughout this time the preferred form of wood fuel was the branches of cut coppice stems bundled into faggots. Larger, bent or deformed stems that were of no other use to the woodland craftsmen were converted to charcoal.

As with most of Europe, these managed woodlands continued to supply their markets right up to the end of World War Two. Since then much of these woodlands have been converted to broad-scale agriculture. Total demand for fuel increased considerably with the industrial revolution but most of this increased demand was met by the new fuel source coal, which was more compact and more suited to the larger scale of the new industries.

During the Edo period of Japan, wood was used for many purposes, and the consumption of wood led Japan to develop a forest management policy during that era. Demand for timber resources was on the rise not only for fuel, but also for construction of ships and buildings, and consequently deforestation was widespread. As a result, forest fires occurred, along with floods and soil erosion. Around 1666, the shōgun made it a policy to reduce logging and increase the planting of trees. This policy decreed that only the shōgun, or a *daimyō*, could authorize the use of wood. By the 18th century, Japan had developed detailed scientific knowledge about silviculture and plantation forestry.

Fireplaces and Stoves

Ceramic stoves are traditional in Northern Europe:
an 18th-century faience stove at Łańcut Castle.

The development of the chimney and the fireplace allowed for more effective exhaustion of the smoke. Masonry heaters or stoves went a step further by capturing much of the heat of the fire and exhaust in a large thermal mass, becoming much more efficient than a fireplace alone.

The metal stove was a technological development concurrent with the industrial revolution. Stoves were manufactured or constructed pieces of equipment that contained the fire on all sides and provided a means for controlling the draft—the amount of air allowed to reach the fire. Stoves have been made of a variety of materials. Cast iron is among the more common. Soapstone (talc), tile, and steel have all been used. Metal stoves are often lined with refractory materials such as firebrick, since the hottest part of a woodburning fire will burn away steel over the course of several years' use.

The Franklin stove was developed in the United States by Benjamin Franklin. More a manufactured fireplace than a stove, it had an open front and a heat exchanger in the back that was designed to draw air from the cellar and heat it before releasing it out the sides. The heat exchanger was never a popular feature and was omitted in later versions. So-called "Franklin" stoves today are made in a great variety of styles, though none resembles the original design.

Potbelly stove at the Museum of Appalachia.

The 1800s became the high point of the cast iron stove. Each local foundry would make their own design, and stoves were built for myriads of purposes—parlour stoves, box stoves, camp stoves, railroad stoves, portable stoves, cooking stoves and so on. Elaborate nickel and chrome edged models took designs to the edge, with cast ornaments, feet and doors. Wood or coal could be burnt in the stoves and thus they were popular for over one hundred years. The action of the fire, combined with the causticity of the ash, ensured that the stove would eventually disintegrate or crack over time. Thus a steady supply of stoves was needed. The maintenance of stoves, needing to be blacked, their smokiness, and the need to split wood meant that oil or electric heat found favour.

The airtight stove, originally made of steel, allowed greater control of combustion, being more tightly fitted than other stoves of the day. Airtight stoves became common in the 19th century.

Use of wood heat declined in popularity with the growing availability of other, less labor-intensive fuels. Wood heat was gradually replaced by coal and later by fuel oil, natural gas and propane heating except in rural areas with available forests.

After the 1967 Oil Embargo, many people in the United States used wood as fuel for the first time. The EPA provided information on clean stoves, which burned much more efficiently.

The growth in popularity of wood heat also led to the development and marketing of a greater variety of equipment for cutting, splitting and processing firewood. Consumer grade hydraulic log splitters were developed to be powered by electricity, gasoline, or PTO of farm tractors. In 1987 the US Department of Agriculture published a method for producing kiln dried firewood, on the basis that better heat output and increased combustion efficiency can be achieved with logs containing lower moisture content.

The magazine "Wood Burning Quarterly" was published for several years before changing its name to "Home Energy Digest" and, subsequently, disappearing.

A pellet stove is an appliance that burns compressed wood or biomass pellets. Wood heat continues to be used in areas where firewood is abundant. For serious attempts at heating, rather than mere ambience (open fireplaces), stoves, fireplace inserts, and furnaces are most commonly used today. In rural, forested parts of the U.S., freestanding boilers are increasingly common. They are installed outdoors, some distance from the house, and connected to a heat exchanger in the house using underground piping. The mess of wood, bark, smoke, and ashes is kept outside and the risk of fire is reduced. The boilers are large enough to hold a fire all night, and can burn larger pieces of wood, so that less cutting and splitting is required. There is no need to retrofit a chimney in the house. However, outdoor wood boilers emit more wood smoke and associated pollutants than other wood-burning appliances. This is due to design characteristics such as the water-filled jacket surrounding the firebox, which acts to cool the fire and leads to incomplete combustion. Outdoor wood boilers also typically have short stack heights in comparison to other wood-burning appliances, contributing to ambient levels of particulates at ground level. An alternative that is increasing in popularity are wood gasification boilers, which burn wood at very high efficiencies (85-91%) and can be placed indoors or in an outbuilding. There are plenty of ways to process wood fuel and the inventions today are maximizing by the minute.

Wood is still used today for cooking in many places, either in a stove or an open fire. It is also used as a fuel in many industrial processes, including smoking meat and making maple syrup.

As a sustainable energy source, wood fuel also remains viable for generating electricity in areas with easy access to forest products and by-products.

Measurement of Firewood

Stapled birch wood.

In the metric system, firewood is normally sold by the cubic metre or stere (1 m³ = ~0.276 cords).

In the United States and Canada, firewood is usually sold by the cord, 128 ft³ (3.62 m³), corresponding to a woodpile 8 ft wide × 4 ft high of 4 ft-long logs. The cord is legally defined by statute in most U.S. states. A "thrown cord" is firewood that has not been stacked and is defined as 4 ft wide x 4 ft tall x 10 ft long. The additional volume is to make it equivalent to a standard stacked cord, where there is less void space. It is also common to see wood sold by the "face cord", which is usually *not* legally defined, and varies from one area to another. For example, in one state a pile of wood 8 feet wide × 4 feet high of 16"-long logs will often be sold as a "face cord", though its volume is only one-third of a cord. In another state, or even another area of the same state, the volume of a face cord may be considerably different. Hence, it is risky to buy wood sold in this manner, as the transaction is not based on a legally enforceable unit of measure.

In Australia, it is normally sold by the tonne but is commonly advertised as sold by the barrowload (wheelbarrow), bucket (1/3 of a m3 bucket of a typical skid-steer), ute-load or bag (roughly 15-20kg).

Energy Content

A common hardwood, red oak, has an energy content (heat value) of 14.9 megajoules per kilogram (6,388 BTU per pound), and 10.4 megajoules recoverable if burned at 70% efficiency.

The Sustainable Energy Development Office (SEDO), part of the Government of Western Australia states that the energy content of wood is 16.2 megajoules per kilogram (4.5 kWh/kg).

According to *The Bioenergy Knowledge Centre*, the energy content of wood is more closely related to its moisture content than its species. The energy content improves as moisture content decreases.

In 2008, wood for fuel cost $15.15 per 1 million BTUs (0.041 EUR per kWh).

Environmental Impacts

Fireplace and chimney after a wildfire, Witch Fire.

Combustion by-products

As with any fire, burning wood fuel creates numerous by-products, some of which may be useful (heat and steam), and others that are undesirable, irritating or dangerous.

One by-product of wood burning is wood ash, which in moderate amounts is a fertilizer (mainly potash), contributing minerals, but is strongly alkaline as it contains potassium hydroxide (lye). Wood ash can also be used to manufacture soap.

Smoke, containing water vapor, carbon dioxide and other chemicals and aerosol particulates, including caustic alkali fly ash, which can be an irritating (and potentially dangerous) by-product of partially burnt wood fuel. A major component of wood smoke is fine particles that may account for a large portion of particulate air pollution in some regions. During cooler months, wood heating accounts for as much as 60% of fine particles in Melbourne, Australia.

Slow combustion stoves increase efficiency of wood heaters burning logs, but also increase particulate production. Low pollution/slow combustion stoves are a current area of research. An alternative approach is to use pyrolysis to produce several useful biochemical byproducts, and clean burning charcoal, or to burn fuel extremely quickly inside a large thermal mass, such as a masonry heater. This has the effect of allowing the fuel to burn completely without producing particulates while maintaining the efficiency of the system.

In some of the most efficient burners, the temperature of the smoke is raised to a much higher temperature where the smoke will itself burn (e.g. 609 °C for igniting carbon monoxide gas). This may result in significant reduction of smoke hazards while also providing additional heat from the process. By using a catalytic converter, the temperature for obtaining cleaner smoke can be reduced. Some U.S. jurisdictions prohibit sale or installation of stoves that do not incorporate catalytic converters.

Combustion by-product Effects on Human Health

Wood-burning fireplace with burning log.

Depending on population density, topography, climatic conditions and combustion equipment used, wood heating may substantially contribute to air pollution, particularly particulates. The conditions in which wood is burnt will greatly influence the content of the emission. Particulate air pollution can contribute to human health problems and increased hospital admissions for asthma & heart diseases.

The technique of compressing wood pulp into pellets or artificial logs can reduce emissions. The combustion is cleaner, and the increased wood density and reduced water content can eliminate some of the transport bulk. The fossil energy consumed in transport is reduced and represents a small fraction of the fossil fuel consumed in producing and distributing heating oil or gas.

Wood combustion products can include toxic and carcinogenic substances. Generally, the heartwood of a tree contains the highest amounts of toxic substances, but precautions should be taken if one is burning wood of an unknown nature, since some trees' woodsmoke can be highly toxic.

Harvesting Operations

Much wood fuel comes from native forests around the world. Plantation wood is rarely used for firewood, as it is more valuable as timber or wood pulp, however, some wood fuel is gathered from trees planted amongst crops, also known as agroforestry. The collection or harvesting of this wood can have serious environmental implications for the collection area. The concerns are often specific to the particular area, but can include all the problems that regular logging create. The heavy removal of wood from forests can cause habitat destruction and soil erosion. However, in many countries, for example in Europe and Canada, the forest residues are being collected and turned into useful wood fuels with minimal impact on the environment. Consideration is given to soil nutrition as well as erosion. The environmental impact of using wood as a fuel depends on how it is burnt. Higher temperatures result in more complete combustion and less noxious gases as a result of pyrolysis. Some may regard the burning of wood from a sustainable source as carbon-neutral. A tree, over the course of its lifetime, absorbs as much carbon (or carbon dioxide) as it releases when burnt.

Some firewood is harvested in "woodlots" managed for that purpose, but in heavily wooded areas it is more often harvested as a byproduct of natural forests. Deadfall that has not started to rot is preferred, since it is already partly seasoned. Standing dead timber is considered better still, as it is both seasoned, and has less rot. Harvesting this form of timber reduces the speed and intensity of bushfires. Harvesting timber for firewood is normally carried out by hand with chainsaws. Thus, longer pieces - requiring less manual labor, and less chainsaw fuel - are less expensive and only limited by the size of their firebox. Prices also vary considerably with the distance from wood lots, and quality of the wood. Firewood usually relates to timber or trees unsuitable for building or construction. Firewood is a renewable resource provided the consumption rate is controlled to sustainable levels. The shortage of suitable firewood in some places has seen local populations damaging huge tracts of bush possibly leading to further desertification.

Greenhouse Gases

Wood burning creates more atmospheric CO_2 than biodegradation of wood in a forest (in a given period of time) because by the time the bark of a dead tree has rotted, the log has already been occupied by other plants and micro-organisms which continue to sequester the CO_2 by integrating the hydrocarbons of the wood into their own life cycle. Wood harvesting and transport operations produce varying degrees of greenhouse gas pollution. Inefficient and incomplete combustion of wood can result in elevated levels of greenhouse gases other than CO_2, which may result in positive emissions where the byproducts have greater Carbon dioxide equivalent values. In an attempt to provide quantitative information about the relative output of CO_2 to produce electricity of domestic heating, the United Kingdom Department of Energy and Climate Change (DECC) has published a comprehensive model comparing the burning of wood (wood chip) and other fuels, based on 33 scenarios. The model's output is kilogram of CO_2 produced per Megawatt hour of delivered energy. Scenario 33 for example, which concerns the production of heat from wood chips produced from UK small roundwood produced from bringing neglected broadleaf forests back into production, shows that burning oil releases 377 kg of CO_2 while burning woodchip releases 1501 kg of CO_2 per MW h delivered energy. On the other hand, scenario 32 in that same reference, which concerns production of heat from wood chips that would otherwise be made into particleboard, releases only 239 kg of CO_2 per MW h delivered energy. Therefore the relative greenhouse effects of biomass energy production very much depends on the usage model.

The intentional and controlled charring of wood and its incorporation into the soil is an effective method for carbon sequestration as well as an important technique to improve soil conditions for agriculture, particularly in heavily forested regions. It forms the basis of the rich soils known as Terra preta.

Regulation and Legislation

The environmental impact of burning wood fuel is debatable. Several cities have moved towards setting standards of use and/or bans of wood burning fireplaces. For example, the city of Montréal, Québec passed a resolution to ban wood fireplace installation in new construction. Wood burning advocates claim that properly harvested wood is carbon-neutral, therefore off-setting the negative impact of by-product particles given off during the burning process. In the context of forest wildfires, wood removed from the forest setting for use as wood fuel can reduce overall emissions by decreasing the quantity of open burned wood and the severity of the burn while combusting the remaining material under regulated conditions. On March 7, 2018, the United States House of Representatives passed a bill that would delay for three years the implementation of more stringent emission standards for new residential wood heaters.

Potential use in Renewable Energy Technologies

Sawmills create and burn sawdust: it can be pelletized and used at home.

- Efficient stove for developing nations

- Pellet stove

- Sawdust can be pelletized

- Wood pellets

Usage

Some European countries produce a significant fraction of their electricity needs from wood or wood wastes. In Scandinavian countries the costs of manual labor to process firewood is very high. Therefore, it is common to import firewood from countries with cheap labor and natural resources. The main exporters to Scandinavia are the Baltic countries (Estonia, Lithuania, and Latvia). In Finland, there is a growing interest in using wood waste as fuel for home and industrial heating, in the form of compacted pellets.

In the United States, wood fuel is the second-leading form of renewable energy (behind hydro-electric).

Australia

About 1.5 million households in Australia use firewood as the main form of domestic heating. As of 1995, approximately 1.85 million cubic metres of firewood (1m³ equals approximately one car trailer load) was used in Victoria annually, with half being consumed in Melbourne. This amount is comparable to the wood consumed by all of Victoria's sawlog and pulplog forestry operations (1.9 million m³).

A pile of firewood logged from the Barmah Forest in Victoria.

Species used as sources of firewood include:

- Red Gum, from forests along the Murray River (the Mid-Murray Forest Management Area, including the Barmah and Gunbower forests, provides about 80% of Victoria's red gum timber).

- Box and Messmate Stringybark, in southern Australia.

- Sugar gum, a wood with high thermal efficiency that usually comes from small plantations.

- Jarrah, in the Southwest of Western Australia. It generates a greater heat than most other available woods and is usually sold by the tonne.

Europe

In 2014, the construction of the biggest pellet plant in the Baltic region was started in Võrumaa, Sõmerpalu, with an expected output of 110,000 tons of pellet / year. Different types of wood will be used in the process of pellet making (firewood, woodchips, shavings). The Warmeston OÜ plant started its activity by the end of 2014. In 2013, the main pellet consumers in Europe were the UK, Denmark, the Netherlands, Sweden, Germany and Belgium, as U.E.'s annual report on biofuels states. In Denmark and Sweden, pellets are used by power plants, households and medium scale consumers for district heating, compared to Austria and Italy, where pellets are mainly used as small - scale private residential and industrial boilers for heating. The UK is the single largest consuming market for industrial wood pellets, in large part due to its major biomass-fueled power stations such as Drax, MGT and Lynemouth.

Asia

Japan and South Korea are both growing markets for industrial wood pellets, and as of 2017, were expected to become the second and third largest global markets for wood pellets due to government policies favoring the use of biomass in power generation.

North America

Demand for wood fuel in the United States is principally driven by residential and commercial heating customers. Canada was not a major consumer of industrial wood pellets as of 2017, but has relatively aggressive de-carbonization policies and may become a significant consumer of industrial wood pellets by the 2020s.

BIOFUEL

Biofuel is any fuel that is derived from biomass—that is, plant or algae material or animal waste. Since such feedstock material can be replenished readily, biofuel is considered to be a source of renewable energy, unlike fossil fuels such as petroleum, coal, and natural gas. Biofuel is commonly advocated as a cost-effective and environmentally benign alternative to petroleum and other fossil fuels, particularly within the context of rising petroleum prices and increased concern over the contributions made by fossil fuels to global warming. Many critics express concerns about the scope of the expansion of certain biofuels because of the economic and environmental costs associated with the refining process and the potential removal of vast areas of arable land from food production.

Ethanol gas fuel pump delivering the E85 mixture to
an automobile in Washington state.

Types of Biofuels

Some long-exploited biofuels, such as wood, can be used directly as a raw material that is burned to produce heat. The heat, in turn, can be used to run generators in a power plant to produce electricity. A number of existing power facilities burn grass, wood, or other kinds of biomass.

Liquid biofuels are of particular interest because of the vast infrastructure already in place to use them, especially for transportation. The liquid biofuel in greatest production is ethanol (ethyl alcohol), which is made by fermenting starch or sugar. Brazil and the United States are among the leading producers of ethanol. In the United States ethanol biofuel is made primarily from corn (maize) grain, and it is typically blended with gasoline to produce "gasohol," a fuel that is 10 percent ethanol. In Brazil, ethanol biofuel is made primarily from sugarcane, and it is commonly used as a 100-percent-ethanol fuel or in gasoline blends containing 85 percent ethanol. Unlike the "first-generation" ethanol biofuel produced from food crops, "second-generation" cellulosic ethanol is derived from low-value biomass that possesses a high cellulose content, including wood chips, crop residues, and municipal waste. Cellulosic ethanol is commonly made from sugarcane bagasse, a waste product from sugar processing, or from various grasses that can be cultivated on low-quality land. Given that the conversion rate is lower than with first-generation biofuels, cellulosic ethanol is dominantly used as a gasoline additive.

An ethanol production plant in South Dakota.

The second most common liquid biofuel is biodiesel, which is made primarily from oily plants (such as the soybean or oil palm) and to a lesser extent from other oily sources (such as waste cooking fat from restaurant deep-frying). Biodiesel, which has found greatest acceptance in Europe, is used in diesel engines and usually blended with petroleum diesel fuel in various percentages. The use of algae and cyanobacteria as a source of "third-generation" biodiesel holds promise but has been difficult to develop economically. Some algal species contain up to 40 percent lipids by weight, which can be converted into biodiesel or synthetic petroleum. Some estimates state that algae and cyanobacteria could yield between 10 and 100 times more fuel per unit area than second-generation biofuels.

Research technician, Nick Sweeney, inoculates algae being grown in a
tent reactor in the algal lab in the Field Test Laboratory Building (FTLB) at the
National Renewable Energy Laboratory in Golden.

Other biofuels include methane gas and biogas—which can be derived from the decomposition of biomass in the absence of oxygen—and methanol, butanol, and dimethyl ether—which are in development.

Economic and Environmental Considerations

In evaluating the economic benefits of biofuels, the energy required to produce them has to be taken into account. For example, the process of growing corn to produce ethanol consumes fossil fuels in farming equipment, in fertilizer manufacturing, in corn transportation, and in ethanol distillation. In this respect, ethanol made from corn represents a relatively small energy gain; the energy gain from sugarcane is greater and that from cellulosic ethanol or algae biodiesel could be even greater.

Biofuels also supply environmental benefits but, depending on how they are manufactured, can also have serious environmental drawbacks. As a renewable energy source, plant-based biofuels in principle make little net contribution to global warming and climate change; the carbon dioxide (a major greenhouse gas) that enters the air during combustion will have been removed from the air earlier as growing plants engage in photosynthesis. Such a material is said to be "carbon neutral." In practice, however, the industrial production of agricultural biofuels can result in additional emissions of greenhouse gases that may offset the benefits of using a renewable fuel. These emissions include carbon dioxide from the burning of fossil fuels during the production process and nitrous oxide from soil that has been treated with nitrogen fertilizer. In this regard, cellulosic biomass is considered to be more beneficial.

Land use is also a major factor in evaluating the benefits of biofuels. The use of regular feedstock, such as corn and soybeans, as a primary component of first-generation biofuels sparked the "food versus fuel" debate. In diverting arable land and feedstock from the human food chain, biofuel production can affect the economics of food price and availability. In addition, energy crops grown for biofuel can compete for the world's natural habitats. For example, emphasis on ethanol derived from corn is shifting grasslands and brushlands to corn monocultures, and emphasis on biodiesel is bringing down ancient tropical forests to make way for oil palm plantations. Loss of natural habitat can change the hydrology, increase erosion, and generally reduce biodiversity of wildlife areas. The clearing of land can also result in the sudden release of a large amount of carbon dioxide as the plant matter that it contains is burned or allowed to decay.

Biofuels testing centre. Workers at the biofuels testing centre at the National Renewable Energy Laboratory (NREL) in Golden.

Some of the disadvantages of biofuels apply mainly to low-diversity biofuel sources—corn, soybeans, sugarcane, oil palms—which are traditional agricultural crops. One alternative involves the use of highly diverse mixtures of species, with the North American tallgrass prairie as a specific example. Converting degraded agricultural land that is out of production to such high-diversity biofuel sources could increase wildlife area, reduce erosion, cleanse waterborne pollutants, store carbon dioxide from the air as carbon compounds in the soil, and ultimately restore fertility to degraded lands. Such biofuels could be burned directly to generate electricity or converted to liquid fuels as technologies develop.

The proper way to grow biofuels to serve all needs simultaneously will continue to be a matter of much experimentation and debate, but the fast growth in biofuel production will likely continue. In the United States the Energy Independence and Security Act of 2007 mandated the use of 136 billion litres (36 billion gallons) of biofuels annually by 2022, more than a sixfold increase over 2006 production levels. The legislation also requires, with certain stipulations, that 79 billion litres (21 billion gallons) of the total amount be biofuels other than corn-derived ethanol, and it continued certain government subsidies and tax incentives for biofuel production.

One distinctive promise of biofuels is that, in combination with an emerging technology called carbon capture and storage, the process of producing and using biofuels may be capable of perpetually removing carbon dioxide from the atmosphere. Under this vision, biofuel crops would remove carbon dioxide from the air as they grow, and energy facilities would capture the carbon dioxide given off as biofuels are burned to generate power. Captured carbon dioxide could be sequestered (stored) in long-term repositories such as geologic formations beneath the land, in sediments of the deep ocean, or conceivably as solids such as carbonates.

References

- Biomass, entry: newworldencyclopedia.org, Retrieved 4 January, 2019

- "Natural Gas – Exports". The World Factbook. Central Intelligence Agency. Retrieved 11 June 2015

- Biomass-renewable, comment: power-technology.com, Retrieved 17 February, 2019

- "Methane Emissions in the Oil and Gas Industry". American Geosciences Institute. 16 May 2018. Retrieved 1 May 2019

- Biomass-home, energyexplained: eia.gov, Retrieved 25 August, 2019

- Minter, George. "socalgas's Minter on Renewable Natural Gas as a Foundational Fuel". Www.planningreport. com. David Abel. Retrieved 3 May 2018

- Biomass-resources, bioenergy: energy.gov, Retrieved 21 April, 2019

- Lee, Jechan; Sarmah, Ajit K.; Kwon, Eilhann E. (2019). Biochar from biomass and waste - Fundamentals and applications. Elsevier. Pp. 1–462. Doi:10.1016/C2016-0-01974-5. Hdl:10344/443. ISBN 978-0-12-811729-3

- Biofuel, technology: britannica.com, Retrieved 15 May, 2019

- Chumpoo, Jade; Prasassarakich, Pattarapan (24 February 2010). "Bio-Oil from Hydro-Liquefaction of Bagasse in Supercritical Ethanol". Energy & Fuels. 24 (3): 2071–2077. Doi:10.1021/ef901241e

- Diamond, Jared. 2005 Collapse: How Societies Choose to Fail or Succeed. Penguin Books. New York. 294–304 pp. ISBN 0-14-303655-6

- Global pellet market outlook in 2017 | Wood Pellet Association of Canada". Www.pellet.org. Retrieved 2018-07-19

Technologies for Bioenergy Production

The important techniques used to derive bioenergy from different sources include biorefinery, bioconversion, thermal depolymerization, Fischer–Tropsch Process, biomass heating system, and carbon capture and storage. All these diverse bioenergy techniques have been carefully analyzed in this chapter.

BIOREFINERY

A biorefinery is a facility that integrates biomass conversion processes and equipment to produce fuels, power, and value-added chemicals from biomass. Biorefinery is analogous to today's petroleum refinery, which produces multiple fuels and products from petroleum. By producing several products, a biorefinery takes advantage of the various components in biomass and their intermediates, therefore maximizing the value derived from the biomass feedstock.

A biorefinery could, for example, produce one or several low-volume, but high-value, chemical products and a low-value, but high-volume liquid transportation fuel such as biodiesel or bioethanol. At the same time, it can generate electricity and process heat, through CHP technology, for its own use and perhaps enough for sale of electricity to the local utility. The high value products increase profitability, the high-volume fuel helps meet energy needs, and the power production helps to lower energy costs and reduce GHG emissions from traditional power plant facilities.

Biorefinery Platforms

There are several platforms which can be employed in a biorefinery with the major ones being the sugar platform and the thermochemical platform (also known as syngas platform).

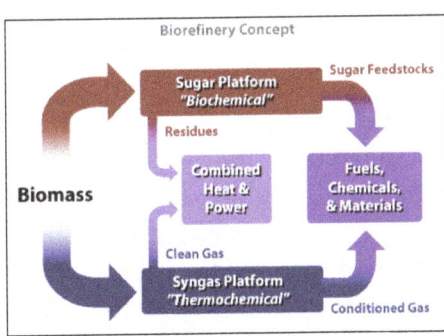

Sugar platform biorefineries breaks down biomass into different types of component sugars for fermentation or other biological processing into various fuels and chemicals. On the other hand, thermochemical biorefineries transform biomass into synthesis gas (hydrogen and carbon monoxide) or pyrolysis oil.

The thermochemical biomass conversion process is complex, and uses components, configurations, and operating conditions that are more typical of petroleum refining. Biomass is converted into syngas, and syngas is converted into an ethanol-rich mixture. However, syngas created from biomass contains contaminants such as tar and sulphur that interfere with the conversion of the syngas into products. These contaminants can be removed by tar-reforming catalysts and catalytic reforming processes. This not only cleans the syngas, it also creates more of it, improving process economics and ultimately cutting the cost of the resulting ethanol.

Advantages

Biorefineries can help in utilizing the optimum energy potential of organic wastes and may also resolve the problems of waste management and GHGs emissions. Biomass wastes can be converted, through appropriate enzymatic/chemical treatment, into either gaseous or liquid fuels. The pre-treatment processes involved in biorefining generate products like paper-pulp, HFCS, solvents, acetate, resins, laminates, adhesives, flavour chemicals, activated carbon, fuel enhancers, undigested sugars etc. which generally remain untapped in the traditional processes. The suitability of this process is further enhanced from the fact that it can utilize a variety of biomass resources, whether plant-derived or animal-derived.

BIOCONVERSION

Bioconversion, also known as *biotransformation*, is the conversion of organic materials, such as plant or animal waste, into usable products or energy sources by biological processes or agents, such as certain microorganisms. One example is the industrial production of cortisone, which one step is the bioconversion of progesterone to 11-alpha-Hydroxyprogesterone by *Rhizopus nigricans*. Another example is the bioconversion of glycerol to 1,3-propanediol, which is part of scientific research for many decades.

Another example of bioconversion is the conversion of organic materials, such as plant or animal waste, into usable products or energy sources by biological processes or agents, such as certain microorganisms, some detritivores or enzymes.

In the USA, the Bioconversion Science and Technology group performs multidisciplinary R&D for the Department of Energy's (DOE) relevant applications of bioprocessing, especially with biomass. Bioprocessing combines the disciplines of chemical engineering, microbiology and biochemistry. The Group 's primary role is investigation of the use of microorganism, microbial consortia and microbial enzymes in bioenergy research. New cellulosic ethanol conversion processes have enabled the variety and volume of feedstock that can be bioconverted to expand rapidly. Feedstock now includes materials derived from plant or animal waste such as paper, auto-fluff, tires, fabric, construction materials, municipal solid waste (MSW), sludge, sewage, etc.

Three Different Processes for Bioconversion

1. Enzymatic hydrolysis - a single source of feedstock, switchgrass for example, is mixed with strong enzymes which convert a portion of cellulosic material into sugars which can then be fermented into ethanol. Genencor and Novozymes are two companies that have received United States government Department of Energy funding for research into reducing the cost of cellulase, a key enzyme in the production cellulosic ethanol by this process.

2. Synthesis gas fermentation - a blend of feedstock, not exceeding 30% water, is gasified in a closed environment into a syngas containing mostly carbon monoxide and hydrogen. The cooled syngas is then converted into usable products through exposure to bacteria or other catalysts. BRI Energy, LLC is a company whose pilot plant in Fayetteville, Arkansas is currently using synthesis gas fermentation to convert a variety of waste into ethanol. After gasification, anaerobic bacteria (*Clostridium ljungdahlii*) are used to convert the syngas (CO, CO_2, and H_2) into ethanol. The heat generated by gasification is also used to co-generate excess electricity.

3. C.O.R.S. and Grub Composting are sustainable technologies that employ organisms that feed on organic matter to reduce and convert organic waste in to high quality feedstuff and oil rich material for the biodiesel industry. Organizations pioneering this novel approach to waste management are EAWAG, ESR International, Prota Culture and BIOCONVERSION that created the *e*-CORS® system to meet large scale organic waste management needs and environmental sustainability in both urban and livestock farming reality. This type of engineered system introduces a substantial innovation represented by the automatic modulation of the treatment, able to adapt conditions of the system to the biology of the scavenger used, improving their performances and the power of this technology.

BIOENERGY WITH CARBON CAPTURE AND STORAGE

Bio-energy with carbon capture and storage (BECCS) is the process of extracting bioenergy from biomass and capturing and storing the carbon, thus removing it from the atmosphere.

The carbon in the trees or crops used for the biomass comes from the greenhouse gas carbon dioxide (CO_2), which they extract from the atmosphere whilst growing. As of 2019 most BECCS stores some of the carbon and leaves the rest with the bioenergy in the form of a biofuel, such as ethanol. By providing aviation biofuel and reducing the environmental impact of aviation BECCS may help with climate change mitigation. Alternatively the bioenergy could be used in the form of electricity and a larger proportion of the CO_2 stored permanently by injection into geological formations.

As of 2019 five facilities around the world are actively using BECCS technologies and are capturing approximately 1.5 million tonnes per year (Mtpa) of CO_2 The IPCC Fifth Assessment Report by the Intergovernmental Panel on Climate Change (IPCC), implies a range of negative emissions from BECCS of 0 to 22 gigatonnes.

There are other non-BECCS forms of carbon dioxide removal and storage including afforestation, biochar, carbon dioxide air capture and biomass burial and enhanced weathering. Pyrogenic carbon capture and storage (PyCCS) or biochar is superior in fixing carbon for a longer time. Carbon dioxide injected into geologic formations eventually leaks back into the atmosphere due to seismic activity and natural faults and problems with the seal of the ancient injection pits.

Negative Emission

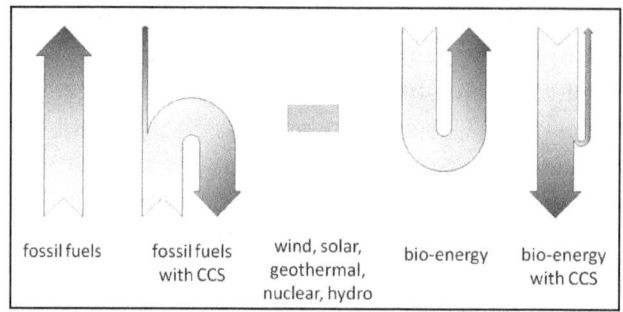

fossil fuels   fossil fuels   wind, solar,   bio-energy   bio-energy
with CCS   geothermal,   with CCS
nuclear, hydro

Carbon flow schematic for different energy systems.

The main appeal of BECCS is in its ability to result in negative emissions of CO_2. The capture of carbon dioxide from bioenergy sources effectively removes CO_2 from the atmosphere.

Bio-energy is derived from biomass which is a renewable energy source and serves as a carbon sink during its growth. During industrial processes, the biomass combusted or processed re-releases the CO_2 into the atmosphere. The process thus results in a net zero emission of CO_2, though this may be positively or negatively altered depending on the carbon emissions associated with biomass growth, transport and processing, Carbon capture and storage (CCS) technology serves to intercept the release of CO_2 into the atmosphere and redirect it into geological storage locations. CO_2 with a biomass origin is not only released from biomass fuelled power plants, but also during the production of pulp used to make paper and in the production of biofuels such as biogas and bioethanol. The BECCS technology can also be employed on such industrial processes.

It is argued that through the BECCS technology, carbon dioxide is trapped in geologic formations for very long periods of time, whereas for example a tree only stores its carbon during its lifetime. In its report on the CCS technology, IPCC projects that more than 99% of carbon dioxide which is stored through geologic sequestration is likely to stay in place for more than 1000 years. While other types of carbon sinks such as the ocean, trees and soil may involve the risk of adverse feedback loops at increased temperatures, BECCS technology is likely to provide a better permanence by storing CO_2 in geological formations.

The amount of CO_2 that has been released to date is believed to be too much to be able to be absorbed by conventional sinks such as trees and soil in order to reach low emission targets. In addition to the presently accumulated emissions, there will be significant additional emissions during this century, even in the most ambitious low-emission scenarios. BECCS has therefore been suggested as a technology to reverse the emission trend and create a global system of net negative emissions. This implies that the emissions would not only be zero, but negative, so that not only the emissions, but the absolute amount of CO_2 in the atmosphere would be reduced.

Application

Source	CO_2 Source	Sector
Ethanol production	Fermentation of biomass such as sugarcane, wheat or corn releases CO_2 as a by-product	Industry
Pulp and paper mills	• CO_2 produced in recovery boilers • CO_2 produced in lime kilns • For gasification technologies, CO_2 is produced during the gasification of black liquor and biomass such as the tree bark and wood. • Huge amounts of CO_2 are also released by the combustion of <u>syngas</u>, a product of gasification, in the combined cycle process.	Industry
Biogas production	In the biogas upgrading process, CO_2 is separated from the methane to produce a higher quality gas	Industry
Electrical power plants	Combustion of biomass or biofuel in steam or gas powered generators releases CO_2 as a by-product	Energy
Heat power plants	Combustion of biofuel for heat generation releases CO_2 as a by-product. Usually used for district heating	Energy

Cost

The IPCC states that estimations for BECCS cost range from \$60-\$250 per ton of CO_2.

Technology

The main technology for CO_2 capture from biotic sources generally employs the same technology as carbon dioxide capture from conventional fossil fuel sources. Broadly, three different types of technologies exist: post-combustion, pre-combustion, and oxy-fuel combustion.

Oxy-combustion

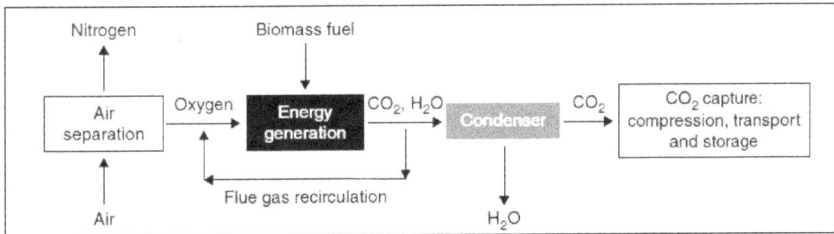

Overview of oxy-fuel combustion for carbon capture from biomass, showing the key processes and stages; some purification is also likely to be required at the dehydration stage.

Oxy-fuel combustion has been a common process in the glass, cement and steel industries. It is also a promising technological approach for CCS. In oxy-fuel combustion, the main difference from conventional air firing is that the fuel is burned in a mixture of O_2 and recycled flue gas. The O_2 is produced by an air separation unit (ASU), which removes the atmospheric N_2 from the oxidizer stream. By removing the N_2 upstream of the process, a flue gas with a high concentration of CO_2 and water vapor is produced, which eliminates the need for a post-combustion capture plant. The water vapor can be removed by condensation, leaving a product stream of relatively high-purity CO_2 which, after subsequent purification and dehydration, can be pumped to a geological storage site.

The key challenges of BECCS implementation using oxy-combustion method is associated with combustion process. For the high volatile content biomass, the mill temperature has to be maintain at low temperature to prevent the risk of fire and explosion. In addition, the flame temperature is lower. Therefore, the concentration of oxygen needs to be increased up to 27-30%.

Pre-combustion

Pre-combustion carbon capture refers to the process of capturing CO_2 before the energy generations. It is often composed of five operating stations: oxygen generation, syngas generation, CO_2 separation, CO_2 compression, and power generation. Basically, the fuel will first go through a gasification process by reacting with oxygen to form a stream of CO and H_2, which is syngas. The products will then go through a water-gas shift reactor to form CO_2 and H_2. The CO_2 that is produced will then be captured, and the H_2, which is a clean source, will be used for combustion to generate energy. The process of gasification combined with syngas production is called Integrated Gasification Combined Cycle (IGCC). Normally, it would require an Air Separation Unit (ASU) to serve as the oxygen source. However, research proves that with the same flue gas, oxygen gasification has little advantage over air gasification, and they both have a similar thermal efficiency of roughly 70% using coal as the fuel source. Thus, the use of ASU is not really necessary in pre-combustion.

Biomass is considered "sulfur-free" as a fuel for the pre-combustion capture. However, there are other trace elements in biomass combustion such as K and Na that could accumulate in the system and finally cause the degradation of the mechanical parts. Thus, further developments of the separation techniques for those trace elements are needed. And also, after the gasification process, CO_2 takes up to 13% - 15.3% by mass in the syngas stream for biomass sources, while it is only 1.7% - 4.4% for coal. This limit the conversion of CO to CO_2 in the water gas shift, and the production rate for H_2 will decrease accordingly. However, the thermal efficiency of the pre-combustion capture using biomass resembles that of coal which is around 62% - 100%. Besides, some researches also proved that instead of using a biomass/water slurry fuel feed, using a dry system is more thermally efficient and also more practical for biomass.

Post-combustion

In addition to pre-combustion and oxy-fuel combustion technologies, post-combustion is a promising technology which can be used to extract CO_2 emission from biomass fuel resources. During the process, CO_2 is separated from the other gases in the flue gas stream after the biomass fuel is burnt and undergo separation process. Because it has the ability to be retrofitted to some existing power plants such as steam boilers or other newly built power stations, post-combustion technology is considered as a better option than pre-combustion technology. According to the fact sheets released in March 2018, the efficiency of post-combustion technology is expected to be 95% while pre-combustion and oxy-combustion capture CO_2 at an efficient rate of 85% and 87.5% respectively.

Development for current post-combustion technologies has not been entirely done due to several problems. One of the major concerns using this technology to capture carbon dioxide is the parasitic energy consumption. If the capacity of the unit is designed to be small, the heat loss to the surrounding is great enough to cause to many negative consequences. Another challenge of post-combustion carbon capture is how to deal with the mixture's components in the flue gases from initial biomass materials after combustion. The mixture consists of a high amount of alkali metals, halogens, acidic elements, and transition metals which might have negative impacts on the efficiency of the process. Thus, the choice of specific solvents and how to manage the solvent process should be carefully designed and operated.

Biomass Feedstocks

Biomass sources used in BECCS include agricultural residues & waste, forestry residue & waste, industrial & municipal wastes, and energy crops specifically grown for use as fuel. Current BECCS projects capture CO_2 from ethanol bio-refinery plants and municipal solid waste (MSW) recycling center.

Up to date, there have been 23 BECCS projects around the world, with the majority in North America and Europe. Today, there are only 6 projects in operation, capturing CO_2 from ethanol bio-refinery plants and MSW recycling centers.

5 BECSS projects have been canceled due to the difficulty of obtaining the permission as well as their economic viability. The canceled projects include: the White Rose CCS Project at Selby, UK can capture about 2 $MtCO_2$/year from Drax power station and store CO_2 at the Bunter Sandstone. The Rufiji Cluster project at Tanzania plan to capture around 5.0-7.0 $MtCO_2$/year and store CO_2 at the Saline Aquifer. The Greenville project at Ohio, USA has capacity of capturing 1 $MtCO_2$/year. The Wallula project was planned to capture 0.75 $MtCO_2$/year at Washington, USA. Finally, the CO_2 Sink project at Ketzin, Germany.

At Ethanol Plants

Illinois Carbon Capture and Storage (IL-CCS) is one of the milestones, being the first industrial-scaled BECCS project, in the early 21st century. Located in Decatur, Illinois, USA, IL-CCS captures CO_2 from Archer Daniels Midland (ADM) ethanol plant. The captured CO_2 is then injected under the deep saline formation at Mount Simon Sandstone. IL-CCS consists of 2 phases. The first being a pilot project which was implemented from 11/2011 to 11/2014. Phase 1 has a capital cost of around 84 million US dollars. Over the 3-year period, the technology successfully captured and sequestered 1million tonne of CO_2 from the ADM plant to the aquifer. No leaking of CO_2 from the injection zone was found during this period. The project is still being monitored for future reference. The success of phase 1 motivated the deployment of phase 2, bringing IL-CCS (and BECCS) to industrial scale. Phase 2 has been in operation since 11/2017 and also use the same injection zone at Mount Simon Sandstone as phase 1. The capital cost for second phase is about 208 million US dollars including 141 million US dollar fund from the Department of Energy. Phase 2 has capturing capacity about 3 time larger than the pilot project (phase 1). Annually, IL-CCS can capture mourned 1 million tonne of CO_2. With the largest of capturing capacity, IL-CCS is currently the largest BECCS project in the world.

In addition to the IL-CCS project, there are about three more projects that capture CO_2 from the ethanol plant at smaller scales. For example, Arkalon at Kansas, USA can capture 0.18-0.29

$MtCO_2$/yr, OCAP at Netherlands can capture about 0.1-0.3 $MtCO_2$/yr, and Husky Energy at Canada can capture 0.09-0.1 $MtCO_2$/yr.

At MSW Recycling Centers

Beside capturing CO_2 from the ethanol plants, currently, there are 2 models in Europe are designed to capture CO_2 from the processing of Municipal Solid Waste. The Klemetsrud Plant at Oslo, Norway use biogenic municipal solid waste to generate 175 GWh and capture 315 Ktonne of CO_2 each year. It uses absorption technology with Aker Solution Advanced Amine solvent as a CO_2 capture unit. Similarly, the ARV Duiven at Netherlands uses the same technology, but it captures less CO_2 than the previous model. ARV Duiven generates around 126 GWh and only capture 50 Ktonne of CO_2 each year.

Techno-economics of BECCS and the TESBiC Project

The largest and most detailed techno-economic assessment of BECCS was carried out by cmcl innovations and the TESBiC group (Techno-Economic Study of Biomass to CCS) in 2012. This project recommended the most promising set of biomass fueled power generation technologies coupled with carbon capture and storage (CCS). The project outcomes lead to a detailed "biomass CCS roadmap" for the U.K.

Challenges

Environmental Considerations

Some of the environmental considerations and other concerns about the widespread implementation of BECCS are similar to those of CCS. However, much of the critique towards CCS is that it may strengthen the dependency on depletable fossil fuels and environmentally invasive coal mining. This is not the case with BECCS, as it relies on renewable biomass. There are however other considerations which involve BECCS and these concerns are related to the possible increased use of biofuels. Biomass production is subject to a range of sustainability constraints, such as: scarcity of arable land and fresh water, loss of biodiversity, competition with food production, deforestation and scarcity of phosphorus. It is important to make sure that biomass is used in a way that maximizes both energy and climate benefits. There has been criticism to some suggested BECCS deployment scenarios, where there would be a very heavy reliance on increased biomass input.

Large areas of land would be required to operate BECCS on an industrial scale. To remove 10 billion tons of CO_2, upwards of 300 million hectares of land area (larger than India) would be required. As a result, BECCS risks using land that could be better suited to agriculture and food production, especially in developing countries.

These systems may have other negative side effects. There is however presently no need to expand the use of biofuels in energy or industry applications to allow for BECCS deployment. There is already today considerable emissions from point sources of biomass derived CO_2, which could be utilized for BECCS. Though, in possible future bio-energy system upscaling scenarios, this may be an important consideration.

Upscaling BECCS would require a sustainable supply of biomass - one that does not challenge our land, water, and food security. Using bio-energy crops as feedstock will not only cause sustainability concerns but also require the use of more fertilizer leading to soil contamination and water pollution. Moreover, crop yield is generally subjected to climate condition, i.e. the supply of this bio-feedstock can be hard to control. Bioenergy sector must also expand to meet the supply level of biomass. Expanding bioenergy would require technical and economic development accordingly.

Technical Challenges

Just as other carbon capture and storage technologies, one of the challenges of applying BECCS technology is to find suitable geographic locations to build combustion plant and to sequester captured CO_2. If biomass sources are not close by the combustion unit, transporting biomass emits CO_2 offsetting the amount of CO_2 captured by BECCS. BECCS also face technical concerns about efficiency of burning biomass. While each type of biomass has a different heating value, biomass in general is a low-quality fuel. Thermal conversion of biomass has a typical efficiency of 20-27%. Coal-fired plant has an efficiency of about 37% for comparison.

BECCS also faces a question whether the process is actually energy positive. Low energy conversion efficiency, energy-intensive biomass supply, combined with the energy required to power the CO_2 capture and storage unit impose energy penalty on the system. This might lead to a low power generation efficiency.

Alternative Biomass Sources

Agricultural and Forestry Residues

Globally, 14 Gt of forestry residue and 4.4 Gt residues from crop production (mainly barley, wheat, corn, sugarcane and rice) are generated every year. This is a significant amount of biomass which can be combusted to generate 26 EJ/year and achieve a 2.8 Gt of negative CO_2 emission through BECCS. Utilizing residues for carbon capture will provide social and economic benefits to rural communities. Using waste from crops and forestry is a way to avoid the ecological and social challenges of BECCS.

Municipal Solid Waste

Municipal solid waste (MSW) is one of the newly developed sources of biomass. Two current BECCS plants are using MSW as feedstocks. Waste collected from daily life is recycled via incineration waste treatment process. Waste goes through high temperature thermal treatment and the heat generated from combusting organic part of waste is used to generate electricity. CO_2 emitted from this process is captured through absorption using MEA. For every 1 kg of waste combusted, 0.7 kg of negative CO_2 emission is achieved. Utilizing solid waste also have other environmental benefits.

Co-firing Coal with Biomass

There are currently 200 cofiring plants in the world, including 40 in the US. Studies showed that by mixing coal with biomass, we could reduce the amount of CO_2 emitted. The concentration of

CO_2 in the flue gas is an important key to determine the efficiency of CO_2 capture technology. The concentration of CO_2 in the flue gas from the co-firing power plant is roughly the same as coal plant, about 15%. This means that we can reduce our reliance on fossil fuel.

Even though co-firing will have some energy penalty, it still offers higher net efficiency than the biomass combustion plants. Co-firing biomass with coal will result more energy production with less input material. Currently, the modern 500 MW coal power plant can take up to 15% biomass without changing the component of the steam boiler. This promising potential allows co-firing power plant become more favorable than dedicated bio-electricity.

It is estimated that by replacing 25% of coal with biomass at existing power plant in China and the U.S, we can reduce emission by 1Gt per year. The amount of negative CO_2 emitted depends on the composition of coal and biomass. 10% biomass can reduce 0.5 Gt CO_2 per year and with 16% biomass can achieve zero emission. Direct-cofiring (20% biomass) give us negative emission of -26 kg CO_2/MWh (from 93 kg CO_2/MWh).

Biomass cofiring with coal has efficiency near those of coal combustion. Cofiring can be easily applied to existing coal-fired power plant at low cost. The implementation of co-firing power plant on the global scale is still a challenge. The biomass resources have to meet strictly the sustainability criteria and the co-firing project would need the support in term of economic and policy from the governments.

Even though co-firing plant may be an immediate contribution to solving the global warming and climate change issues, co-firing still has some challenges that need to consider. Due to the moisture content of biomass, it will affect the calorific value of the combustor. In addition, high volatile biomass would highly influence the reaction rate and the temperature of the reactor; especially, it may lead to the explosion of furnace.

Instead of co-firing, full conversion from coal to biomass of one or more generating units in a plant may be preferred.

Policy

Based on the current Kyoto Protocol agreement, carbon capture and storage projects are not applicable as an emission reduction tool to be used for the Clean Development Mechanism (CDM) or for Joint Implementation (JI) projects. Recognising CCS technologies as an emission reduction tool is vital for the implementation of such plants as there is no other financial motivation for the implementation of such systems. There has been growing support to have fossil CCS and BECCS included in the protocol. Accounting studies on how this can be implemented, including BECCS, have also been done.

European Union

There are some future policies that give incentives to use bioenergy such as Renewable Energy Directive (RED) and Fuel Quality Directive (FQD), which require 20% of total energy consumption to be based on biomass, bioliquids and biogas by 2020.

United Kingdom

In 2018 the Committee on Climate Change recommended that aviation biofuels should provide up to 10% of total aviation fuel demand by 2050, and that all aviation biofuels should be produced with CCS as soon as the technology is available.

United States

In February, 2018, US congress significantly increased and extended the section 45Q tax credit for sequestration of carbon oxides. This has been a top priority of carbon capture and sequestration (CCS) supporters for several years. It increased $25.70 to $50 tax credit per tonnes of CO_2 for secure geological storage and $15.30 to $35 tax credit per tonne of CO_2 used in enhanced oil recovery.

THERMAL DEPOLYMERIZATION

Thermal depolymerization is an industrial process of breaking down various waste materials into crude oil products. The materials are subjected to high temperatures and pressure in the presence of water, thereby initiating hydrous pyrolysis. As a result, the long chain polymers of the materials are depolymerized into short chain monomers. It is said to mimic the natural geological processes thought to be involved in fossil fuel production.

Thermal depolymerization occurs in nature when an accumulated biomass is heated and pressurized in the earth's crust over millions of years. This biomass, also known as kerogen, is believed to react with clay mineral enzymes at temperatures below 200 °C (392 °F), which produces oil. This method is rapidly gaining a lot of attention world-wide as an alternative source of energy. It is particularly helpful as solid wastes contain carbon, which can be chemically transformed into liquid fuel.

Thermal Depolymerization Process

During thermal depolymerization process, the feedstock material is ground into tiny chunks and mixed with water. The mixture is then subjected to high pressure and heated at a constant volume to 250 °C (482 °F). As a result, crude hydrocarbons and solid minerals are produced, which are then separated using fractional distillation and oil refining techniques. Some of commonly used feedstock materials include corn, soya, sugarcane, tires, sewage sludge and medical wastes. Carthage plant products like aromatics, olefins, paraffins and naphthenes are also used.

The following are the three main steps involved in the thermal depolymerization process:

- Feedstock is heated under pressure and pulped into a water slurry.

- Slurry is subjected to low pressure and then oil is separated from water.

- Crude oil is heated to high temperature to obtain light carbons in a solid form.

The temperature of the initial phase will be in the range of 200 to 300 °C (392 to 572 °F) and the next phase will be around 500 °C (932 °F).

Benefits of Thermal Depolymerization

Thermal depolymerization process can breakdown organic poisons by breaking the chemical bonds and deforming the molecular shape required for the poison's activity. It can also eliminate heavy metals from the samples by converting the metals from their ionized forms to stable oxides that can be separated from the other products.

Using this process, the energy content of organic materials can be recycled without removing the water. Unlike other recovering methods like pyrolysis and burning, which require pre-drying or produce gaseous products, water is easily separated by liquid fuel in this method. TDP energy farms can also be used as a habitat for other species and as recreational space for people.

Limitations of Thermal Depolymerization

Thermal depolymerization process only breaks long molecular chains into short chains. As a result, small molecules like methane or carbon dioxide cannot be converted into oil using this process. Hence, there is a need for additional refining steps. In addition, as the process requires temperature greater than 400 °C (752 °F), toxic byproducts like furan and dioxin may be released in addition to methane and carbon dioxide.

Applications

The key applications of thermal depolymerization are as follows:

- Waste Reduction - Thermal depolymerization is a high heat process that involves physical and chemical changes resulting in total destruction of waste bringing a significant reduction in the volume and mass of the waste. It can treat a wide range of waste including biological, anatomical, plastics, glass and needles, etc.

- Oil Production – Researchers have produced oil from agricultural plant wastes like hog manure, animal wastes, plastics using thermal depolymerization method where the application of heat and pressure yields oil in addition to carbon dioxide, methane and water.

Environmental Impact

Besides reducing waste and by-products by using water as a medium, thermal depolymerization process also produces fuel resources that can benefit the world. Invaluable fuel products can be produced from organic waste and low quality feed stocks in an environment-friendly manner. It also yields clean crude oil products by removing sulfur and nitrogen compounds.

In a completely thermal depolymerization-based economy, the amount of CO_2 produced by the burning of fuels is exactly balanced by the plants grown to be used for thermal depolymerization feedstock. The amount of energy hitting the Earth is about 5000 times more than the total amount of energy used by all human activity. Therefore, with an optimum use of the thermal depolymerization technology, the Earth might conveniently support ten times its current population at a high standard of living.

The technology is yet to make an impact as an alternative for producing liquid crude oil due to high cost of the entire process. Upon resolving this issue, thermal depolymerization could help bring the shortage of the dwindling fuel resources across the globe to an end.

FISCHER–TROPSCH PROCESS

The Fischer–Tropsch process is a collection of chemical reactions that converts a mixture of carbon monoxide and hydrogen into liquid hydrocarbons. These reactions occur in the presence of metal catalysts, typically at temperatures of 150–300 °C (302–572 °F) and pressures of one to several tens of atmospheres. The process was first developed by Franz Fischer and Hans Tropsch at the Kaiser-Wilhelm-Institut für Kohlenforschung in Mülheim an der Ruhr, Germany, in 1925.

As a premier example of C1 chemistry, the Fischer–Tropsch process is an important reaction in both coal liquefaction and gas to liquids technology for producing liquid hydrocarbons. In the usual implementation, carbon monoxide and hydrogen, the feedstocks for FT, are produced from coal, natural gas, or biomass in a process known as gasification. The Fischer–Tropsch process then converts these gases into a synthetic lubrication oil and synthetic fuel. The Fischer–Tropsch process has received intermittent attention as a source of low-sulfur diesel fuel and to address the supply or cost of petroleum-derived hydrocarbons.

Reaction Mechanism

The Fischer–Tropsch process involves a series of chemical reactions that produce a variety of hydrocarbons, ideally having the formula (C_nH_{2n+2}). The more useful reactions produce alkanes as follows:

$$(2n + 1)\,H_2 + n\,CO \rightarrow C_nH_{2n+2} + n\,H_2O$$

where n is typically 10–20. The formation of methane ($n = 1$) is unwanted. Most of the alkanes produced tend to be straight-chain, suitable as diesel fuel. In addition to alkane formation, competing reactions give small amounts of alkenes, as well as alcohols and other oxygenated hydrocarbons.

Fischer–Tropsch Intermediates and Elemental Reactions

Converting a mixture of H_2 and CO into aliphatic products is a multi-step reaction with several intermediate compounds. The growth of the hydrocarbon chain may be visualized as involving a repeated sequence in which hydrogen atoms are added to carbon and oxygen, the C–O bond is split and a new C–C bond is formed. For one $-CH_2-$ group produced by $CO + 2\,H_2 \rightarrow (CH_2) + H_2O$, several reactions are necessary:

- Associative adsorption of CO.

- Splitting of the C–O bond.

- Dissociative adsorption of 2 H_2.

- Transfer of 2 H to the oxygen to yield H_2O.

- Desorption of H_2O.

- Transfer of 2 H to the carbon to yield CH_2.

The conversion of CO to alkanes involves hydrogenation of CO, the hydrogenolysis (cleavage with H_2) of C–O bonds, and the formation of C–C bonds. Such reactions are assumed to proceed via initial formation of surface-bound metal carbonyls. The CO ligand is speculated to undergo dissociation, possibly into oxide and carbide ligands. Other potential intermediates are various C_1 fragments including formyl (CHO), hydroxycarbene (HCOH), hydroxymethyl (CH_2OH), methyl (CH_3), methylene (CH_2), methylidyne (CH), and hydroxymethylidyne (COH). Furthermore, and critical to the production of liquid fuels, are reactions that form C–C bonds, such as migratory insertion. Many related stoichiometric reactions have been simulated on discrete metal clusters, but homogeneous Fischer–Tropsch catalysts are poorly developed and of no commercial importance.

Addition of isotopically labelled alcohol to the feed stream results in incorporation of alcohols into product. This observation establishes the facility of C–O bond scission. Using ^{14}C-labelled ethylene and propene over cobalt catalysts results in incorporation of these olefins into the growing chain. Chain growth reaction thus appears to involve both 'olefin insertion' as well as 'CO-insertion'.

Feedstocks: Gasification

Fischer–Tropsch plants associated with coal or related solid feedstocks (sources of carbon) must first convert the solid fuel into gaseous reactants, *i.e.*, CO, H_2, and alkanes. This conversion is called gasification and the product is called synthesis gas ("syngas"). Synthesis gas obtained from coal gasification tends to have a H_2:CO ratio of ~0.7 compared to the ideal ratio of ~2. This ratio is adjusted via the water-gas shift reaction. Coal-based Fischer–Tropsch plants produce varying amounts of CO_2, depending upon the energy source of the gasification process. However, most coal-based plants rely on the feed coal to supply all the energy requirements of the Fischer–Tropsch process.

Feedstocks: GTL

Carbon monoxide for FT catalysis is derived from hydrocarbons. In gas to liquids (GTL) technology, the hydrocarbons are low molecular weight materials that often would be discarded or flared. Stranded gas provides relatively cheap gas. GTL is viable provided gas remains relatively cheaper than oil.

Several reactions are required to obtain the gaseous reactants required for Fischer–Tropsch catalysis. First, reactant gases entering a Fischer–Tropsch reactor must be desulfurized. Otherwise, sulfur-containing impurities deactivate ("poison") the catalysts required for Fischer–Tropsch reactions.

Several reactions are employed to adjust the H_2:CO ratio. Most important is the water-gas shift reaction, which provides a source of hydrogen at the expense of carbon monoxide:

$$H_2O + CO \rightarrow H_2 + CO_2$$

For Fischer–Tropsch plants that use methane as the feedstock, another important reaction is steam reforming, which converts the methane into CO and H_2:

$$H_2O + CH_4 \rightarrow CO + 3\,H_2$$

Process Conditions

Generally, the Fischer–Tropsch process is operated in the temperature range of 150–300 °C (302–572 °F). Higher temperatures lead to faster reactions and higher conversion rates but also tend to favor methane production. For this reason, the temperature is usually maintained at the low to middle part of the range. Increasing the pressure leads to higher conversion rates and also favors formation of long-chained alkanes, both of which are desirable. Typical pressures range from one to several tens of atmospheres. Even higher pressures would be favorable, but the benefits may not justify the additional costs of high-pressure equipment, and higher pressures can lead to catalyst deactivation via coke formation.

A variety of synthesis-gas compositions can be used. For cobalt-based catalysts the optimal H_2:CO ratio is around 1.8–2.1. Iron-based catalysts can tolerate lower ratios, due to intrinsic water-gas shift reaction activity of the iron catalyst. This reactivity can be important for synthesis gas derived from coal or biomass, which tend to have relatively low H_2:CO ratios (< 1).

Design of the Fischer–Tropsch Process Reactor

Efficient removal of heat from the reactor is the basic need of Fischer–Tropsch reactors since these reactions are characterized by high exothermicity. Four types of reactors are discussed:

- Multi Tubular Fixed-bed Reactor: This type of reactor contains a number of tubes with small diameter. These tubes contain catalyst and are surrounded by boiling water which removes the heat of reaction. A fixed-bed reactor is suitable for operation at low temperatures and has an upper temperature limit of 257 °C (530 K). Excess temperature leads to carbon deposition and hence blockage of the reactor. Since large amounts of the products formed are in liquid state, this type of reactor can also be referred to as a trickle flow reactor system.

- Entrained Flow Reactor: An important requirement of the reactor for the Fischer–Tropsch process is to remove the heat of the reaction. This type of reactor contains two banks of heat exchangers which remove heat; the remainder of which is removed by the products and recycled in the system. The formation of heavy waxes should be avoided, since they condense on the catalyst and form agglomerations. This leads to fluidization. Hence, risers are operated over 297 °C (570 K).

- Slurry Reactors: Heat removal is done by internal cooling coils. The synthesis gas is bubbled through the waxy products and finely-divided catalyst which is suspended in the liquid medium. This also provides agitation of the contents of the reactor. The catalyst particle size reduces diffusional heat and mass transfer limitations. A lower temperature in the reactor leads to a more viscous product and a higher temperature (> 297 °C, 570 K) gives an undesirable product spectrum. Also, separation of the product from the catalyst is a problem.

- Fluid-bed and circulating catalyst (riser) reactors: These are used for high-temperature Fischer–Tropsch synthesis (nearly 340 °C) to produce low-molecular-weight unsaturated hydrocarbons on alkalized fused iron catalysts. The fluid-bed technology (as adapted from the catalytic cracking of heavy petroleum distillates) was introduced by Hydrocarbon

Research in 1946–50 and named the 'Hydrocol' process. A large scale Fischer–Tropsch Hydrocol plant (350,000 tons per annum) operated during 1951–57 in Brownsville, Texas. Due to technical problems, and lacking economy due to increasing petroleum availability, this development was discontinued. Fluid-bed Fischer–Tropsch synthesis has recently been very successfully reinvestigated by Sasol. One reactor with a capacity of 500,000 tons per annum is now in operation and even larger ones are being built (nearly 850,000 tons per annum). The process is now used mainly for C_2 and C_7 alkene production. This new development can be regarded as an important progress in Fischer–Tropsch technology.

A high-temperature process with a circulating iron catalyst ('circulating fluid bed', 'riser reactor', 'entrained catalyst process') was introduced by the Kellogg Company and a respective plant built at Sasol in 1956. It was improved by Sasol for successful operation. At Secunda, South Africa, Sasol operated 16 advanced reactors of this type with a capacity of approximately 330,000 tons per annum each. Now the circulating catalyst process is being replaced by the superior Sasol-advanced fluid-bed technology. Early experiments with cobalt catalyst particles suspended in oil have been performed by Fischer. The bubble column reactor with a powdered iron slurry catalyst and a CO-rich syngas was particularly developed to pilot plant scale by Kölbel at the Rheinpreuben Company in 1953. Recently (since 1990) low-temperature Fischer–Tropsch slurry processes are under investigation for the use of iron and cobalt catalysts, particularly for the production of a hydrocarbon wax, or to be hydrocracked and isomerised to produce diesel fuel, by Exxon and Sasol. Today slurry-phase (bubble column) low-temperature Fischer–Tropsch synthesis is regarded by many authors as the most efficient process for Fischer–Tropsch clean diesel production. This Fischer–Tropsch technology is also under development by the Statoil Company (Norway) for use on a vessel to convert associated gas at offshore oil fields into a hydrocarbon liquid.

Product Distribution

In general the product distribution of hydrocarbons formed during the Fischer–Tropsch process follows an Anderson–Schulz–Flory distribution, which can be expressed as:

$$\frac{W_n}{n} = (1 - \alpha)^2 \alpha^{n-1}$$

where W_n is the weight fraction of hydrocarbons containing n carbon atoms, and α is the chain growth probability or the probability that a molecule will continue reacting to form a longer chain. In general, α is largely determined by the catalyst and the specific process conditions.

Examination of the above equation reveals that methane will always be the largest single product so long as α is less than 0.5; however, by increasing α close to one, the total amount of methane formed can be minimized compared to the sum of all of the various long-chained products. Increasing α increases the formation of long-chained hydrocarbons. The very long-chained hydrocarbons are waxes, which are solid at room temperature. Therefore, for production of liquid transportation fuels it may be necessary to crack some of the Fischer–Tropsch products. In order to avoid this, some researchers have proposed using zeolites or other catalyst substrates with fixed sized pores that can restrict the formation of hydrocarbons longer than some characteristic size (usually $n < 10$). This way they can drive the reaction so as to minimize methane formation without producing lots of long-chained hydrocarbons. Such efforts have had only limited success.

Catalysts

A variety of catalysts can be used for the Fischer–Tropsch process, the most common are the transition metals cobalt, iron, and ruthenium. Nickel can also be used, but tends to favor methane formation ("methanation").

Cobalt

Cobalt-based catalysts are highly active, although iron may be more suitable for certain applications. Cobalt catalysts are more active for Fischer–Tropsch synthesis when the feedstock is natural gas. Natural gas has a high hydrogen to carbon ratio, so the water-gas shift is not needed for cobalt catalysts. Iron catalysts are preferred for lower quality feedstocks such as coal or biomass. Synthesis gases derived from these hydrogen-poor feedstocks has a low-hydrogen-content and require the water-gas shift reaction. Unlike the other metals used for this process (Co, Ni, Ru), which remain in the metallic state during synthesis, iron catalysts tend to form a number of phases, including various oxides and carbides during the reaction. Control of these phase transformations can be important in maintaining catalytic activity and preventing breakdown of the catalyst particles.

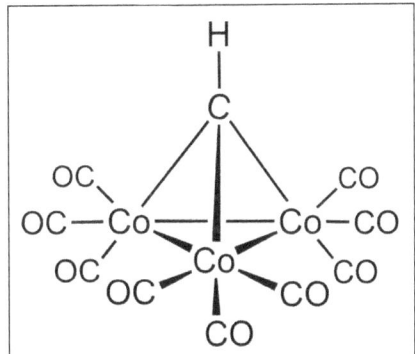

Methylidynetricobaltnonacarbonyl is a molecule that illustrates the kind
of reduced carbon species speculated to occur in the Fischer–Tropsch process.

In addition to the active metal the catalysts typically contain a number of "promoters," including potassium and copper. Group 1 alkali metals, including potassium, are a poison for cobalt catalysts but are promoters for iron catalysts. Catalysts are supported on high-surface-area binders/supports such as silica, alumina, or zeolites. Promotors also have an important influence on activity. Alkali metal oxides and copper are common promotors, but the formulation depends on the primary metal, iron vs cobalt. Alkali oxides on cobalt catalysts generally cause activity to drop severely even with very low alkali loadings. $C_{\geq 5}$ and CO_2 selectivity increase while methane and C_2–C_4 selectivity decrease. In addition, the alkene to alkane ratio increases.

Fischer–Tropsch catalysts are sensitive to poisoning by sulfur-containing compounds. Cobalt-based catalysts are more sensitive than their iron counterparts.

Iron

Fischer–Tropsch iron catalysts need alkali promotion to attain high activity and stability (e.g. 0.5 wt% K_2O Addition of Cu for reduction promotion, addition of SiO_2, Al_2O_3 for structural promotion and maybe some manganese can be applied for selectivity control (e.g. high olefinicity). The working catalyst is only obtained when—after reduction with hydrogen—in the initial period of synthesis several iron carbide phases and elemental carbon are formed whereas iron oxides are still present in addition to some metallic iron. With iron catalysts two directions of selectivity have been pursued. One direction has aimed at a low-molecular-weight olefinic hydrocarbon mixture to be produced in an entrained phase or fluid bed process (Sasol–Synthol process). Due to the relatively high reaction temperature (approx. 340 °C), the average molecular weight of the product is so low that no liquid product phase occurs under reaction conditions. The catalyst particles moving around in the reactor are small (particle diameter 100 μm) and carbon deposition on the catalyst does not disturb reactor operation. Thus a low catalyst porosity with small pore diameters as obtained from fused magnetite (plus promoters) after reduction with hydrogen is appropriate. For maximising the overall gasoline yield, C_3 and C_4 alkenes have been oligomerized at Sasol. However, recovering the olefins for use as chemicals in, e.g., polymerization processes is advantageous today. The second direction of iron catalyst development has aimed at highest catalyst activity to be used at low reaction temperature where most of the hydrocarbon product is in the liquid phase under reaction conditions. Typically, such catalysts are obtained through precipitation from nitrate solutions. A high content of a carrier provides mechanical strength and wide pores for easy mass transfer of the reactants in the liquid product filling the pores. The main product fraction then is a paraffin wax, which is refined to marketable wax materials at Sasol; however, it also can be very selectively hydrocracked to a high quality diesel fuel. Thus, iron catalysts are very flexible.

Ruthenium

Ruthenium is the most active of the FT catalysts. It works at the lowest reaction temperatures, and it produces the highest molecular weight hydrocarbons. It acts as a Fischer–Tropsch catalyst as the pure metal, without any promotors, thus providing the simplest catalytic system of Fischer–Tropsch synthesis, where mechanistic conclusions should be the easiest—e.g., much easier than with iron as the catalyst. Like with nickel, the selectivity changes to mainly methane at elevated temperature. Its high price and limited world resources exclude industrial application. Systematic Fischer–Tropsch studies with ruthenium catalysts should contribute substantially to the further exploration of the fundamentals of Fischer–Tropsch synthesis. There is an interesting question to consider: what features have the metals nickel, iron, cobalt, and ruthenium in common to let them—and only them—be Fischer–Tropsch catalysts, converting the CO/H_2 mixture to aliphatic (long chain) hydrocarbons in a 'one step reaction'. The term 'one step reaction' means that reaction intermediates are not desorbed from the catalyst surface. In particular, it is amazing that the much carbided alkalized iron catalyst gives a similar reaction as the just metallic ruthenium catalyst.

HTFT and LTFT

High-Temperature Fischer–Tropsch (or HTFT) is operated at temperatures of 330–350 °C and uses an iron-based catalyst. This process was used extensively by Sasol in their coal-to-liquid plants (CTL). Low-Temperature Fischer–Tropsch (LTFT) is operated at lower temperatures and uses an iron or cobalt-based catalyst. This process is best known for being used in the first integrated GTL-plant operated and built by Shell in Bintulu, Malaysia.

Commercialization

Ras Laffan, Qatar

ORYX GTL Plant – Qatar.

The LTFT facility Pearl GTL at Ras Laffan, Qatar, is the largest FT plant. It uses cobalt catalysts at 230 °C, converting natural gas to petroleum liquids at a rate of 140,000 barrels per day (22,000 m³/d), with additional production of 120,000 barrels (19,000 m³) of oil equivalent in natural gas liquids and ethane. The plant in Ras Laffan was commissioned in 2007, called Oryx GTL, has a capacity of 34,000 barrels per day (5,400 m³/d). The plant utilizes the Sasol slurry phase distillate process, which uses a cobalt catalyst. Oryx GTL is a joint venture between Qatar Petroleum and Sasol.

Sasol

A SASOL garage in Gauteng.

Another large scale implementation of Fischer–Tropsch technology is a series of plants operated by Sasol in South Africa, a country with large coal reserves, but little oil. The first commercial plant opened in 1952. Sasol uses coal and now natural gas as feedstocks and produces a variety of synthetic petroleum products, including most of the country's diesel fuel.

Sasol scrapped plans to build the GTL plant in Westlake, Louisiana.

PetroSA

PetroSA, another South African company, operates a refinery with a 36,000 barrels a day plant that completed semi-commercial demonstration in 2011, paving the way to begin commercial preparation. The technology can be used to convert natural gas, biomass or coal into synthetic fuels.

Shell Middle Distillate Synthesis

One of the largest implementations of Fischer–Tropsch technology is in Bintulu, Malaysia. This Shell facility converts natural gas into low-sulfur Diesel fuels and food-grade wax. The scale is 12,000 barrels per day (1,900 m³/d).

Velocys

Construction is underway for Velocys' commercial reference plant incorporating its microchannel Fischer–Tropsch technology; ENVIA Energy's Oklahoma City GTL project being built adjacent to Waste Management's East Oak landfill site. The project is being financed by a joint venture between Waste Management, NRG Energy, Ventech and Velocys. The feedstock for this plant will be a combination of landfill gas and pipeline natural gas.

UPM (Finland)

In October 2006, Finnish paper and pulp manufacturer UPM announced its plans to produce biodiesel by the Fischer–Tropsch process alongside the manufacturing processes at its European paper and pulp plants, using waste biomass resulting from paper and pulp manufacturing processes as source material.

Rentech

A demonstration-scale Fischer–Tropsch plant was built and operated by Rentech, Inc., in partnership with ClearFuels, a company specializing in biomass gasification. Located in Commerce City, Colorado, the facility produces about 10 barrels per day (1.6 m³/d) of fuels from natural gas. Commercial-scale facilities are planned for Rialto, California; Natchez, Mississippi; Port St. Joe, Florida; and White River, Ontario. Rentech closed down their pilot plant in 2013, and abandoned work on their FT process as well as the proposed commercial facilities.

INFRA GTL Technology

In 2010, INFRA built a compact Pilot Plant for conversion of natural gas into synthetic oil. The plant modeled the full cycle of the GTL chemical process including the intake of pipeline gas, sulfur removal, steam methane reforming, syngas conditioning, and Fischer–Tropsch synthesis. In 2013 the first pilot plant was acquired by VNIIGAZ Gazprom LLC. In 2014 INFRA commissioned and operated on a continuous basis a new, larger scale full cycle Pilot Plant. It represents the second generation of INFRA's testing facility and is differentiated by a high degree of automation and extensive data gathering system. In 2015, INFRA built its own catalyst factory in Troitsk (Moscow, Russia). The catalyst factory has a capacity of over 15 tons per year, and produces the unique proprietary Fischer–Tropsch catalysts developed by the company's R&D division. In 2016, INFRA designed and built a modular, transportable GTL (gas-to-liquid) M100 plant for processing natural and associated gas into synthetic crude oil in Wharton (Texas, USA). The M100 plant is operating as a technology demonstration unit, R&D platform for catalyst refinement, and economic model to scale the Infra GTL process into larger and more efficient plants.

Other

In the United States and India, some coal-producing states have invested in Fischer–Tropsch plants. In Pennsylvania, Waste Management and Processors, Inc. was funded by the state to implement Fischer–Tropsch technology licensed from Shell and Sasol to convert so-called waste coal (leftovers from the mining process) into low-sulfur diesel fuel.

Research Developments

Choren Industries has built a plant in Germany that converts biomass to syngas and fuels using the Shell Fischer–Tropsch process structure. The company went bankrupt in 2011 due to impracticalities in the process.

Biomass gasification (BG) and Fischer–Tropsch (FT) synthesis can in principle be combined to produce renewable transportation fuels (biofuels).

U.S. Air Force Certification

Syntroleum, a publicly traded United States company, has produced over 400,000 U.S. gallons (1,500,000 L) of diesel and jet fuel from the Fischer–Tropsch process using natural gas and coal at its demonstration plant near Tulsa, Oklahoma. Syntroleum is working to commercialize its licensed Fischer–Tropsch technology via coal-to-liquid plants in the United States, China, and Germany, as well as gas-to-liquid plants internationally. Using natural gas as a feedstock, the ultra-clean, low sulfur fuel has been tested extensively by the United States Department of Energy (DOE) and the United States Department of Transportation (DOT). Most recently, Syntroleum has been working with the United States Air Force to develop a synthetic jet fuel blend that will help the Air Force to reduce its dependence on imported petroleum. The Air Force, which is the United States military's largest user of fuel, began exploring alternative fuel sources in 1999. On December 15, 2006, a B-52 took off from Edwards Air Force Base, California for the first time powered solely by a 50–50 blend of JP-8 and Syntroleum's FT fuel. The seven-hour flight test was considered a success. The goal of the flight test program is to qualify the fuel blend for fleet use on the service's B-52s, and then flight test and qualification on other aircraft. The test program concluded in 2007. This program is part of the Department of Defense Assured Fuel Initiative, an effort to develop secure domestic sources for the military energy needs. The Pentagon hopes to reduce its use of crude oil from foreign producers and obtain about half of its aviation fuel from alternative sources by 2016. With the B-52 now approved to use the FT blend, the C-17 Globemaster III, the B-1B, and eventually every airframe in its inventory to use the fuel by 2011.

Carbon Dioxide Reuse

Carbon dioxide is not a typical feedstock for FT catalysis. Hydrogen and carbon dioxide react over a cobalt-based catalyst, producing methane. With iron-based catalysts unsaturated short-chain hydrocarbons are also produced. Upon introduction to the catalyst's support, ceria functions as a reverse water-gas shift catalyst, further increasing the yield of the reaction. The short-chain hydrocarbons were upgraded to liquid fuels over solid acid catalysts, such as zeolites.

Process Efficiency

Using conventional FT technology the process ranges in carbon efficiency from 25 to 50 percent and a thermal efficiency of about 50% for CTL facilities idealised at 60% with GTL facilities at about 60% efficiency idealised to 80% efficiency.

Fischer–Tropsch in Nature

A Fischer–Tropsch-type process has also been suggested to have produced a few of the building blocks of DNA and RNA within asteroids. Similarly, the hypothetical abiogenic petroleum formation requires some naturally occurring FT-like processes.

BIOCONVERSION OF BIOMASS TO MIXED ALCOHOL FUELS

The bioconversion of biomass to mixed alcohol fuels can be accomplished using the MixAlco process. Through bioconversion of biomass to a mixed alcohol fuel, more energy from the biomass will end up as liquid fuels than in converting biomass to ethanol by yeast fermentation.

The process involves a biological/chemical method for converting any biodegradable material (e.g., urban wastes, such as municipal solid waste, biodegradable waste, and sewage sludge, agricultural residues such as corn stover, sugarcane bagasse, cotton gin trash, manure) into useful chemicals, such as carboxylic acids (e.g., acetic, propionic, butyric acid), ketones (e.g., acetone, methyl ethyl ketone, diethyl ketone) and biofuels, such as a mixture of primary alcohols (e.g., ethanol, propanol, *n*-butanol) and/or a mixture of secondary alcohols (e.g., isopropanol, 2-butanol, 3-pentanol). Because of the many products that can be economically produced, this process is a true biorefinery.

Pilot Plant.

The process uses a mixed culture of naturally occurring microorganisms found in natural habitats such as the rumen of cattle, termite guts, and marine and terrestrial swamps to anaerobically digest biomass into a mixture of carboxylic acids produced during the acidogenic and acetogenic stages of anaerobic digestion, however with the inhibition of the methanogenic final stage. The more popular methods for production of ethanol and cellulosic ethanol use enzymes that must be isolated first to be added to the biomass and thus convert the starch or cellulose into simple sugars, followed then by yeast fermentation into ethanol. This process does not need the addition of such enzymes as these microorganisms make their own.

As the microoganisms anaerobically digest the biomass and convert it into a mixture of carboxylic acids, the pH must be controlled. This is done by the addition of a buffering agent (e.g., ammonium bicarbonate, calcium carbonate), thus yielding a mixture of carboxylate salts. Methanogenesis, being the natural final stage of anaerobic digestion, is inhibited by the presence of the ammonium ions or by the addition of an inhibitor (e.g., iodoform). The resulting fermentation broth contains the produced carboxylate salts that must be dewatered. This is achieved efficiently by vapor-compression evaporation. Further chemical refining of the dewatered fermentation broth may then take place depending on the final chemical or biofuel product desired.

The condensed distilled water from the vapor-compression evaporation system is recycled back to the fermentation. On the other hand, if raw sewage or other waste water with high BOD in need of treatment is used as the water for the fermentation, the condensed distilled water from the evaporation can be recycled back to the city or to the original source of the high-BOD waste water. Thus, this process can also serve as a water treatment facility, while producing valuable chemicals or biofuels.

Because the system uses a mixed culture of microorganisms, besides not needing any enzyme addition, the fermentation requires no sterility or aseptic conditions, making this front step in the process more economical than in more popular methods for the production of cellulosic ethanol. These savings in the front end of the process, where volumes are large, allows flexibility for further chemical transformations after dewatering, where volumes are small.

Carboxylic Acids

Carboxylic acids can be regenerated from the carboxylate salts using a process known as "acid springing". This process makes use of a high-molecular-weight tertiary amine (e.g., trioctylamine), which is switched with the cation (e.g., ammonium or calcium). The resulting amine carboxylate can then be thermally decomposed into the amine itself, which is recycled, and the corresponding carboxylic acid. In this way, theoretically, no chemicals are consumed or wastes produced during this step.

Ketones

There are two methods for making ketones. The first one consists on thermally converting calcium carboxylate salts into the corresponding ketones. This was a common method for making acetone from calcium acetate during World War I. The other method for making ketones consists on converting the vaporized carboxylic acids on a catalytic bed of zirconium oxide.

Alcohols

Primary Alcohols

The undigested residue from the fermentation may be used in gasification to make hydrogen (H_2). This H_2 can then be used to hydrogenolyze the esters over a catalyst (e.g., copper chromite), which are produced by esterifying either the ammonium carboxylate salts (e.g., ammonium acetate, propionate, butyrate) or the carboxylic acids (e.g., acetic, propionic, butyric acid) with a high-molecular-weight alcohol (e.g., hexanol, heptanol). From the hydrogenolysis, the final products are the high-molecular-weight alcohol, which is recycled back to the esterification, and the corresponding primary alcohols (e.g., ethanol, propanol, butanol).

Secondary Alcohols

The secondary alcohols (e.g., isopropanol, 2-butanol, 3-pentanol) are obtained by hydrogenating over a catalyst (e.g., Raney nickel) the corresponding ketones (e.g., acetone, methyl ethyl ketone, diethyl ketone).

Drop-in Biofuels

The primary or secondary alcohols obtained as described above may undergo conversion to drop-in biofuels, fuels which are compatible with current fossil fuel infrastructure such as biogasoline, green diesel and bio-jet fuel. Such is done by subjecting the alcohols to dehydration followed by oligomerization using zeolite catalysts in a manner similar to the methanex process, which used to produce gasoline from methanol in New Zealand.

Acetic Acid versus Ethanol

Cellulosic-ethanol manufacturing plants are bound to be net exporters of electricity because a large portion of the lignocellulosic biomass, namely lignin, remains undigested and it must be burned, thus producing electricity for the plant and excess electricity for the grid. As the market grows and this technology becomes more widespread, coupling the liquid fuel and the electricity markets will become more and more difficult.

Acetic acid, unlike ethanol, is biologically produced from simple sugars without the production of carbon dioxide:

$$C_6H_{12}O_6 \rightarrow 2\ CH_3CH_2OH + 2\ CO_2$$

(Biological production of ethanol)

$$C_6H_{12}O_6 \rightarrow 3\ CH_3COOH$$

(Biological production of acetic acid)

Because of this, on a mass basis, the yields will be higher than in ethanol fermentation. If then, the undigested residue (mostly lignin) is used to produce hydrogen by gasification, it is ensured that more energy from the biomass will end up as liquid fuels rather than excess heat/electricity.

$$3\ CH_3COOH + 6\ H_2 \rightarrow 3\ CH_3CH_2OH + 3\ H_2O$$

(Hydrogenation of acetic acid)

$$C_6H_{12}O_6 \text{ (from cellulose)} + 6\ H_2 \text{ (from lignin)} \rightarrow 3\ CH_3CH_2OH + 3\ H_2O$$

(Overall reaction)

A more comprehensive description of the economics of each of the fuels is given on the pages alcohol fuel and ethanol fuel, more information about the economics of various systems can be found on the central page biofuel.

Stage of Development

The system has been in development since 1991, moving from the laboratory scale (10 g/day) to the pilot scale (200 lb/day) in 2001. A small demonstration-scale plant (5 ton/day) has been constructed and is under operation and a 220 ton/day demonstration plant is expected in 2012.

BIOMASS HEATING SYSTEM

Biomass heating systems generate heat from biomass. The systems fall under the categories of:

- Direct combustion

- Gasification

- Combined heat and power (CHP)

- Anaerobic digestion

- Aerobic digestion

Wood chips in a storage hopper, in the middle an agitator to transport the material with a screw conveyor to the boiler.

Benefits of Biomass Heating

The use of biomass in heating systems is beneficial because it uses agricultural, forest, urban and industrial residues and waste to produce heat and/or electricity with less effect on the environment than fossil fuels. This type of energy production has a limited long-term effect on the environment because the carbon in biomass is part of the natural carbon cycle; while the carbon in fossil fuels is not, and permanently adds carbon to the environment when burned for fuel (carbon footprint). Historically, before the use of fossil fuels in significant quantities, biomass in the form of wood fuel provided most of humanity's heating.

Forest Health

Because forest based biomass is typically derived from wood that has lower commercial value, forest biomass is typically harvested as a byproduct of other timber harvest operations. Biomass heating provides markets for lower value wood, which enables healthy and profitable forest management. In New England as of 2017, one of the greatest threats to forest health is conversion from forest to agriculture and development. Harvard Forest scientists reported in 2017, that 65 acres of forest were being lost per day by conversion. By providing markets for low grade wood, the value of forests is enhanced, which makes conversion to housing or agriculture less likely.

Drawbacks of Biomass Heating

On a large scale, the use of agricultural biomass removes agricultural land from food production, reduces the carbon sequestration capacity of forests that are not managed sustainably, and extracts nutrients from the soil. Combustion of biomass creates air pollutants and adds significant quantities of carbon to the atmosphere that may not be returned to the soil for many decades. The time delay between when biomass is burned and the time when carbon is pulled from the atmosphere as a plant or tree grows to replace it is known as carbon debt. The concept of carbon debt is subject to debate. Actual carbon impacts may be subject to philosophy, scale of harvest, land type, biomass type (grass, maize, new wood, waste wood, algae, for example), soil type, and other factors.

Using biomass as a fuel produces air pollution in the form of carbon monoxide, NOx (nitrogen oxides), VOCs (volatile organic compounds), particulates and other pollutants, in some cases at levels above those from traditional fuel sources such as coal or natural gas. Black carbon – a pollutant created by incomplete combustion of fossil fuels, biofuels, and biomass – is possibly the second largest contributor to global warming. In 2009 a Swedish study of the giant brown haze that periodically covers large areas in South Asia determined that it had been principally produced by biomass burning, and to a lesser extent by fossil-fuel burning. Researchers measured a significant concentration of ^{14}C, which is associated with recent plant life rather than with fossil fuels. Modern biomass burning appliances dramatically reduce harmful emissions with advanced technology such as oxygen trim systems.

On combustion, the carbon from biomass is released into the atmosphere as carbon dioxide (CO_2). The amount of carbon stored in dry wood is approximately 50% by weight. When from agricultural sources, plant matter used as a fuel can be replaced by planting for new growth. When the biomass is from forests, the time to recapture the carbon stored is generally longer, and the carbon storage capacity of the forest may be reduced overall if destructive forestry techniques are employed.

The forest biomass-is-carbon-neutral proposal put forward in the early 1990s has been superseded by more recent science that recognizes that mature, intact forests sequester carbon more effectively than cut-over areas. When a tree's carbon is released into the atmosphere in a single pulse, it contributes to climate change much more than woodland timber rotting slowly over decades. Some studies indicate that "even after 50 years the forest has not recovered to its initial carbon storage" and "the optimal strategy is likely to be protection of the standing forest". Other studies show that carbon storage is dependent upon the forest and the use of the harvested biomass. Forests are often managed for multiple aged trees with more frequent, smaller harvests of mature trees. These forests interact with carbon differently than mature forests that are clear-cut. Also, the more efficient the conversion of wood to energy, the less wood that is used and shorter the carbon cycle will be.

Biomass Heating in our World

Biomass heating system for one building complex in the Spanish Basque Country.

The oil price increases since 2003 and consequent price increases for natural gas and coal have increased the value of biomass for heat generation. Forest renderings, agricultural waste, and crops grown specifically for energy production become competitive as the prices of energy dense fossil fuels rise. Efforts to develop this potential may have the effect of regenerating mismanaged croplands and be a cog in the wheel of a decentralized, multi-dimensional renewable energy industry. Efforts to promote and advance these methods became common throughout the European Union through the 2000s. In other areas of the world, inefficient and polluting means to generate heat from biomass coupled with poor forest practices have significantly added to environmental degradation.

Buffer Tanks

Buffer tanks store the hot water the biomass appliance generates and circulates it around the heating system. Sometimes referred to as 'thermal stores', they are crucial for the efficient operation of all biomass boilers where the system loading fluctuates rapidly, or the volume of water in the complete hydraulic system is relatively small. Using a suitably sized buffer vessel prevents rapid cycling of the boiler when the loading is below the minimum boiler output. Rapid cycling of the boiler causes a large increase in harmful emissions such as Carbon monoxide, dust, and NOx, greatly reduces boiler efficiency and increases electrical consumption of the unit. In addition, service and maintenance requirements will be increased as parts are stressed by rapid heating and cooling cycles. Although most boilers claim to be able to turn down to 30% of nominal output, in the real world this is often not achievable due to differences in the fuel from the 'ideal' or test fuel. A suitably sized buffer tank should therefore be considered where the loading of the boiler drops below 50% of the nominal output – in other words unless the biomass component is purely base load, the system should include a buffer tank. In any case where the secondary system does not contain sufficient water for safe removal of residual heat from the biomass boiler irrespective of the loading conditions, the system must include a suitably sized buffer tank. The residual heat from a biomass unit varies greatly depending on the boiler design and the thermal mass of the combustion chamber. Light weight, fast response boilers require only 10L/kW, while industrial wet wood units with very high thermal mass require 40L/kW.

Types of Biomass Heating Systems

Biomass heating plant in Austria; the heat power is about 1000 kW.

The use of Biomass in heating systems has a use in many different types of buildings, and all have different uses. There are four main types of heating systems that use biomass to heat a boiler. The types are Fully Automated, Semi-Automated, Pellet-Fired, and Combined Heat and Power.

Fully Automated

In fully automated systems chipped or ground up waste wood is brought to the site by delivery trucks and dropped into a holding tank. A system of conveyors then transports the wood from the holding tank to the boiler at a certain managed rate. This rate is managed by computer controls and a laser that measures the load of fuel the conveyor is bringing in. The system automatically goes on and off to maintain the pressure and temperature within the boiler. Fully automated systems offer a great deal of ease in their operation because they only require the operator of the system to control the computer, and not the transport of wood while offering comprehensive and cost effective solutions to complex industrial challenges.

Semi-automated or "Surge Bin"

Semi-automated or "Surge Bin" systems are very similar to fully automated systems except they require more manpower to keep operational. They have smaller holding tanks, and a much simpler conveyor systems which will require personnel to maintain the systems operation. The reasoning for the changes from the fully automated system is the efficiency of the system. The heat created by the combustor can be used to directly heat the air or it can be used to heat water in a boiler system which acts as the medium by which the heat is delivered. Wood fire fueled boilers are most efficient when they are running at their highest capacity, and the heat required most days of the year will not be the peak heat requirement for the year. Considering that the system will only need to run at a high capacity a few days of the year, it is made to meet the requirements for the majority of the year to maintain its high efficiency.

Pellet-fired

The third main type of biomass heating systems are pellet-fired systems. Pellets are a processed form of wood, which make them more expensive. Although they are more expensive, they are much more condensed and uniform, and therefore are more efficient. Further, it is relatively easy to automatically feed pellets to boilers. In these systems, the pellets are stored in a grain-type storage silo, and gravity is used to move them to the boiler. The storage requirements are much smaller for pellet-fired systems because of their condensed nature, which also helps cut down costs. these systems are used for a wide variety of facilities, but they are most efficient and cost effective for places where space for storage and conveyor systems is limited, and where the pellets are made fairly close to the facility.

Agricultural Pellet Systems

One subcategory of pellet systems are boilers or burners capable of burning pellet with higher ash rate (paper pellets, hay pellets, straw pellets). One of this kind is PETROJET pellet burner with rotating cylindrical burning chamber. In terms of efficiencies advanced pellet boilers can exceed other forms of biomass because of the more stable fuel charataristics. Advanced pellet boilers can even work in condensing mode and cool down combustion gases to 30-40 °C, instead of 120 °C before sent into the flue.

Combined Heat and Power

Combined heat and power systems are very useful systems in which wood waste, such as wood chips, is used to generate power, and heat is created as a byproduct of the power generation system. They have a very high cost because of the high pressure operation. Because of this high pressure operation, the need for a highly trained operator is mandatory, and will raise the cost of operation. Another drawback is that while they produce electricity they will produce heat, and if producing heat is not desirable for certain parts of the year, the addition of a cooling tower is necessary, and will also raise the cost.

There are certain situations where CHP is a good option. Wood product manufacturers would use a combined heat and power system because they have a large supply of waste wood, and a need for both heat and power. Other places where these systems would be optimal are hospitals and prisons, which need energy, and heat for hot water. These systems are sized so that they will produce enough heat to match the average heat load so that no additional heat is needed, and a cooling tower is not needed.

PELLET STOVE

A pellet stove is a stove that burns compressed wood or biomass pellets to create a source of heat for residential and sometimes industrial spaces. By steadily feeding fuel from a storage container (hopper) into a burn pot area, it produces a constant flame that requires little to no physical adjustments. Today's central heating systems operated with wood pellets as a renewable energy source can reach an efficiency factor of more than 90%.

Method

The pellet fuel is delivered from the storage facility or the day tank (single stoves) into the combustion chamber. With the heat generated, circuit water is heated in the pellet boiler. In central heating systems the hot water then runs through the heating circuit. The heat distribution is the same as other central heating systems. Unlike oil or gas heating, the inclusion of a hot water reservoir is recommended with pellet heating systems to save hot water until it is needed.

Benefits

Pellet-burning central heating system in basement of family home.

Most pellet stoves are self-igniting and cycle themselves on and off under thermostatic control. Stoves with automatic ignition can be equipped with remote controls. Recent innovations include integrated microcontroller monitoring of various safety conditions and can run diagnostic tests if an imminent problem arises.

A properly cleaned and maintained pellet stove should not create creosote, the sticky, flammable substance that causes chimney fires. Pellets burn very cleanly and create only a layer of fine fly ash as a byproduct of combustion. The grade of pellet fuel affects the performance and ash output. Premium-grade pellets produce less than one percent ash content, while standard or low grade pellets produce up to six percent ash. Pellet stove users should be aware of the extra maintenance required with a lower grade pellet, and that inconsistent wood quality can cause serious effects to the electronic machinery over a short period of time.

A pellet stove is normally associated with pelletized wood. However, many pellet stoves will also burn fuels such as grain, corn, seeds, or woodchips. In some pellet stoves, these fuels may need to be mixed with wood pellets. Pelletized trash (containing mostly waste paper) is also a fuel for pellet stoves.

Unlike wood stoves, which operate exclusively on a principle of chimney draft, a pellet stove must use specially sealed exhaust pipe to prevent exhaust gases escaping into the living space due to the air pressure produced by a combustion blower. Pellet stoves require certified double-walled venting, normally three or four inches in diameter with a stainless steel interior and galvanized exterior. Because pellet stoves have a forced exhaust system, they have the advantage of not always requiring a vertical rise to vent, although a 3-to-5-foot (0.91 to 1.52 m) vertical run to induce some draft is recommended to prevent leakage in the case of a power outage. Like a modern gas appliance, pellet stoves can be vented horizontally through an outside wall and terminated below the roof line, making it an excellent choice for structures without an existing chimney. If an existing chimney is available, manufacturers urge use of a correctly sized stainless steel liner the length of the chimney for proper drafting. Modern building techniques have created tightly sealed homes, forcing many pellet stove manufacturers to recommend their stoves be installed with outside air intake to ensure the stoves will run efficiently and prevent potential negative pressure within the home.

Pellet stoves are approved for use in mobile homes, while standard wood-burning stoves are not.

In many states pellet fuel is exempt from sales tax.

Tax Credit

Until January 1, 2012, in most states in the U.S., a 75% efficient pellet stove was eligible for a tax credit up to 30% of the cost of the appliance as part of the 25C provision, plus labor.

Principles of Operation

A pellet stove burn pot.

A pellet stove normally consists of these components, whether basic or complex:

- A hopper.

- An auger system.

- Two blower fans: combustion and convection.

- A firebox: burn pot and ash collection system, sometimes lined with ceramic fiber panels.

- Various safety features (vacuum switch, heat sensors).

- A controller.

To properly function, a pellet stove uses electricity and can be connected to a standard electrical outlet. A pellet stove, like an automatic coal stoker, is a consistent heater consuming fuel that is fed evenly from a refillable hopper into the burn pot (a perforated cast-iron or steel basin), through a motorized system. The most commonly used distributor is an auger system that consists of a spiral length of metal encased in a tube. This mechanism is either located above the burn pot or slightly beneath and guides a portion of pellet fuel from the hopper upwards until it falls into the burn pot for combustion.

Fan systems are necessary for clean, economical performance. The flame produced is concentrated and intense in the small area of the burn pot as a combustion blower introduces air into the bottom of the burn pot, while also forcing exhaust gases into the chimney. While some pellet stoves will be hot to the touch (especially on the viewing window), most manufacturers utilize a series of cast-iron or steel heat exchangers that run along the back and top areas of the visible firebox. With a convection blower, room air is circulated through the heat exchangers and directed into the living space. This method allows for a much higher efficiency than the radiant heat of a hand-fed wood or coal stove, and will in most cases cause the top, sides, and back of the stove to be at most warm to the touch. Along with convection air, an exhaust fan forces air from the firebox through special venting specifically made for pellet fuel. This cycle of circulation is an integral part of the combustion system as well, for the concentrated high temperature flame will quickly overheat the firebox. The possible problems associated with overheating are electrical component failure and flames traveling into the auger tube causing a hopper fire. As safeguards, all pellet stoves are equipped with heat sensors, and sometimes vacuum sensors, enabling the controller to shut down if an unsafe condition is detected. For daily maintenance, an ash vacuum is recommended. These are similar to shop vacs, but are designed for the removal of ash materials. These vacuums are available with a pellet stove kit which enables the cleaning of the interior areas of the stove which improves efficiency.

Pellet stoves can either be lit manually or through an automatic igniter. The igniter piece resembles a car's electric cigarette lighter heating coil. Most models have automatic ignition and can be readily equipped with thermostats or remote controls.

Pellet stoves are routinely tested in laboratory for
improved performances and different fuels.

Corn Stove

A corn stove is designed for whole-kernel shelled corn kernel combustion and is similar to a pellet stove. The chief difference between a pellet stove and a dedicated corn stove is the addition of metal stirring rod within the burnpot or an active ash removal system. These vary in design slightly, but usually consist of one long metal stalk with smaller rods welded at a perpendicular angle, in order to churn the burn pot as it spins. An active ash removal system consists of augers at the bottom of the burn pot that evacuate the ash and clinkers. During a normal burn cycle, the sugar content within corn (and other similar bio-fuels) will cause the ashes to stick together, forming a hard mass. The metal stirring rod breaks apart these masses, causing a much more consistent burn. While there is demand to create stoves that are able to burn multiple fuels with minimal adjustments, some pellet stoves are not designed to stir fuel and cannot burn corn fuel.

References

- Biorefinery: bioenergyconsult.com, Retrieved 21 July, 2019

- Diener, Stefan; Zurbrügg, Christian; Tockner, Klement (2009-06-05). "Conversion of organic material by black soldier fly larvae: establishing optimal feeding rates". Waste Management & Research. SAGE Publications. 27 (6): 603–610. Doi:10.1177/0734242x09103838. ISSN 0734-242X

- "U.S. Product Supplied for Crude Oil and Petroleum Products". Tonto.eia.doe.gov. Archived from the original on 28 February 2011. Retrieved 3 April 2018

- Obersteiner, M. (2001). "Managing Climate Risk". Science. 294(5543): 786–7. Doi:10.1126/sci-ence.294.5543.786b. PMID 11681318

- "Wildlands & Woodlands | Harvard Forest". Harvardforest.fas.harvard.edu. Retrieved 15 May 2019

- G. Cassman, Kenneth; Liska, Adam J. (2007). "Food and fuel for all: Realistic or foolish?". Biofuels, Bioproducts and Biorefining. 1: 18–23. Doi:10.1002/bbb.3

- 2009 State of the World, Into a Warming World,Worldwatch Institute, 56–57, ISBN 978-0-393-33418-0

Bioenergy Feedstocks

The biological materials which can be used directly as a fuel or converted to another form of fuel or energy product are called feedstocks. They can be classified as starch-based feedstocks, oilseed-based feedstocks, fiber and grass cellulosic feedstocks and algae-based feedstocks. This chapter has been carefully written to provide an easy understanding of these types of bioenergy feedstocks.

STARCH-BASED FEEDSTOCKS

Corn

Corn (Zea mays) is a popular feedstock for ethanol production in the United States due to its abundance and relative ease of conversion to ethyl alcohol (ethanol). Corn and other high-starch grains have been converted into ethanol for thousands of years, yet only in the past century has its use as fuel greatly expanded. Conversion includes grinding, cooking with enzymes, fermentation with yeast, and distillation to remove water. For fuel ethanol, two more steps are included: using a molecular sieve to remove the last of the water and denaturing to make the ethanol undrinkable.

Current Potential for use as a Biofuel

Corn grain makes a good biofuel feedstock due to its starch content and its comparatively easy conversion to ethanol. Infrastructure to plant, harvest, and store corn in mass quantities benefits the corn ethanol industry. Unlike sugarcane, in which squeezed sugar water can be directly fermented, corn starch must be cooked with alpha and gluco-amylase enzymes to convert the starch to simple sugars. Cellulosic feedstocks are even more recalcitrant and require time and energy to convert to simple sugars. Under the renewable fuel standard set by Congress in 2007 (RFS-2), grain-based ethanol can make up 15 billion gallons of the 36 billion gallon-per-year requirement. Corn-based ethanol production capacity in 2009 was 10.6 billion gallons. The addition of idled capacity would increase potential production to 12.5 billion gallons per year.

Corn production in the United States reached record highs in 2009 with 13.2 billion bushels from 86.5 million acres. Using the current corn-to-ethanol conversion of 2.8 gallons of ethanol from a bushel of corn, total U.S. corn production could result in approximately 37 billion gallons of ethanol, which would provide approximately 26% of our 137 billion gallon-per-year gasoline consumption. However, using all of our corn for ethanol is neither realistic nor necessary and has not been proposed. Creating the 15 billion gallons required under the RFS-2 would call for 5.4 billion bushels or about 41% of our 2009 corn crop. Although this percentage seems rather high, one-third of the weight and 100% of the nutritional content of corn entering an ethanol dry mill biorefinery is returned to the feed market as distillers grains. These distillers grains can be used to replace corn in the diets of cattle, swine, and poultry. When this replacement is calculated into the overall consumption figures, it lowers the number to 27% of our 2009 corn crop, or only 3.6 billion bushels of corn to produce the 15 billion required gallons. Throughout history, the United States has seen a steady increase in the yields of both corn and ethanol production. It is very likely that the United States will be able to increase corn ethanol production without expanding to new acres and still have plenty of corn remaining to meet other domestic use and export demands.

One bushel of corn grain will yield one-third ethanol, one-third distillers grains, and one-third carbon dioxide, or 17 pounds of distillers grains and 2.8 gallons of ethanol.

Biology and Adaptation

Corn originated in Central America with the first domestication, purported to be in the Tehuacan Valley of Mexico. Spreading throughout the North American continent, corn became an important crop for early Americans. At its peak in 1917, 111 million acres of corn were planted in the United States. Today corn is planted on every continent in the world except Antarctica and is grown throughout many states in the United States, ranging from southern North Dakota to Texas and eastward to New York. Corn is well adapted to growing in temperatures between 50° and 86°Fahrenheit. To produce grain, corn will use approximately 22 to 28 inches of water, which requires 12 to 20 inches of rainfall or irrigation during the growing season. Despite popular misconceptions, nearly 90% of U.S.-grown corn is fed by natural rainfall only, with no irrigation necessary. Many parts of the upper Midwest are well suited to grow corn, and this area is sometimes referred to as the Corn Belt.

Production and Agronomic Information

Corn is seeded between March and May and harvested between September and November in most years. A majority of corn planted today has genetic resistance to some weeds, insects, and plant pathogens. Corn hybrid resistance to various pests and pathogens is a result of biotechnology and plant breeding. Biotechnology traits aid producers in the control of weeds and insects, greatly reducing the amount of pesticides entering the environment. Much of the Corn Belt rotates with other crops, such as soybeans or wheat, to break weed, insect, and disease cycles as well as to reduce the cost of production. Corn responds best to highly fertile soils with supplemental fertilizer applied in most years. Fertilizer may be inorganic chemical fertilizer or manure.

Major nutrients required by corn are nitrogen, phosphorus, and potassium. Inorganic nitrogen fertilizer production is very energy-intensive and as a result, nitrogen fertilizer represents nearly

30% of the energy inputs in corn production. Other major inputs include diesel fuel for tractors, transportation, irrigation, and electricity for irrigation and grain storage.

Potential Yields

The average national corn yield was 165 bushels per acre in 2009. Corn yields have increased by approximately 2 bushels per acre each year since 1940. This increase will likely continue into the future, with some predicting the yield trend will amplify at a greater rate due to biotechnology and advancements in breeding. Ethanol yield per acre would be 462 gallons per acre from corn yields of 165 bushels per acre. An acre of sugarcane can produce an approximate 35 ton yield, resulting in about 560 gallons of sugarcane ethanol.

Production Challenges

Corn production has been blessed with nearly 100 years of infrastructure build-up and research. Farmers have great knowledge and experience in growing corn. This infrastructure and grower intelligence make corn a natural crop for expanded uses such as ethanol. Yet high production costs and high inputs make corn a very intensive crop. Other bioenergy crops may be less intensive and require fewer inputs. The cost versus profit per acre needs to be compared, as economics is a major driver in deciding which crop is best. Growing another crop on an acre where corn could be grown carries risks that may include a new cropping system; no harvest, transport, or storage infrastructure; or no commodity market to fall back on if the biofuel market fails.

Estimated Production Costs

Production costs vary widely depending on tillage, irrigation, yield goal (soil fertility), spraying schedule, seed selection, and rotation. A sample corn budget with rain-fed, no-till, biotech seed, corn/soybean rotation, and 120 bushel yield goal would include a total cost of $211 per acre. If overhead – crop insurance, land, taxes – is included, the total is $305 per acre. Production costs increase to over $600 on irrigated fields with continuous corn.

Environmental and Sustainability Issues

Life cycle analysis (LCA) of ethanol production from corn grain has yielded a net energy ratio of 1.2 to 1.45, which represents just a 20% to 45% positive energy balance in producing ethanol from corn. A major criticism of corn ethanol has been the large amount of fossil energy used in production.

Environmental issues in corn production revolve around erosion, pesticide use, and nutrient use. Pesticides and nutrients have the potential to contaminate surface and ground water. Soil erosion has led to loss of topsoil and polluted streams and river systems with silt. Continuous attention to these issues has led to improvements, yet they will remain concerns in crop production.

Sweet Potato

Sweet potato represents an important biomass resource for fuel alcohol production, because of its chemical composition and high density of starch, compared to other forms of biomass, and thus premise as an alternative bioresource for the production of ethanol through fermentation. Sweet potato is a tropical and temperate regions' crop, normally found in Indian sub-continent. It contains starch (178 g /kg), total sugars (26 g /kg) and protein (3.2 g/kg) on fresh weight basis. The starch can be hydrolysed to monomer units of carbohydrates and can be used by the microorganisms in fermentation process.

Cultivation

The plant does not tolerate frost. It grows best at an average temperature of 24 °C (75 °F), abundant sunshine and warm nights. Annual rainfalls of 750–1,000 mm (30–39 in) are considered most suitable, with a minimum of 500 mm (20 in) in the growing season. The crop is sensitive to drought at the tuber initiation stage 50–60 days after planting, and it is not tolerant to water-logging, as it may cause tuber rots and reduce growth of storage roots if aeration is poor.

Depending on the cultivar and conditions, tuberous roots mature in two to nine months. With care, early-maturing cultivars can be grown as an annual summer crop in temperate areas, such as the Eastern United States and China. Sweet potatoes rarely flower when the daylight is longer than 11 hours, as is normal outside of the tropics. They are mostly propagated by stem or root cuttings or by adventitious shoots called "slips" that grow out from the tuberous roots during storage. True seeds are used for breeding only.

They grow well in many farming conditions and have few natural enemies; pesticides are rarely needed. Sweet potatoes are grown on a variety of soils, but well-drained, light- and medium-textured soils with a pH range of 4.5–7.0 are more favorable for the plant. They can be grown in poor soils with little fertilizer. However, sweet potatoes are very sensitive to aluminum toxicity and will die about six weeks after planting if lime is not applied at planting in this type of soil. Because they are sown by vine cuttings rather than seeds, sweet potatoes are relatively easy to plant. Because the rapidly growing vines shade out weeds, little weeding is needed. A commonly used herbicide to rid the soil of any unwelcome plants that may interfere with growth is DCPA, also known as Dacthal. In the tropics, the crop can be maintained in the ground and harvested as needed for market or home consumption. In temperate regions, sweet potatoes are most often grown on larger farms and are harvested before first frosts.

In the Southeastern United States, sweet potatoes are traditionally cured to improve storage, flavor, and nutrition, and to allow wounds on the periderm of the harvested root to heal. Proper curing requires drying the freshly dug roots on the ground for two to three hours, then storage at 29–32 °C (85–90 °F) with 90 to 95% relative humidity from five to fourteen days. Cured sweet potatoes can keep for thirteen months when stored at 13–15 °C (55–59 °F) with >90% relative humidity. Colder temperatures injure the roots.

Production and Yield

In 2016, global production of sweet potatoes was 105 million tonnes, led by China with 67% of the world total.

In 2016, the world average annual yield for sweet potato crop was 13 tonnes per hectare. The most productive yield of sweet potatoes was in Senegal, where the nationwide average annual yield was 39 tonnes per hectare.

Sweet potatoes with different skin colors.

Sweet potatoes are cultivated throughout tropical and warm temperate regions wherever there is sufficient water to support their growth. Sweet potatoes became common as a food crop in the islands of the Pacific Ocean, South India, Uganda and other African countries.

Cassava

Cassava (Manihot esculenta Crantz) is a mostly vegetatively propagated perennial root crop that grows well in tropical climates. Nevertheless, the roots (main reason for growing cassava) are very perishable once taken from the soil and go to waste unless processed in some way soon after harvest. Most processing requires removal of peels (cortex and periderm), head, and tail ends. These components usually discarded as waste, engender environmental pollution. The components are referred to collectively as cassava peeling residues (CPR), and instead of being discarded as waste, would be put to bioenergy production function. The CPR is generated during production of numerous cassava root based food products like akpakpuru, attieke, casabe, chickwangue, farina (farinha de mandioca), fufu, fuku, gaplek, gari, ijapu, konkonte, lafun, landang, peujeum, and thundam. Because more than 65% of global annual cassava output is processed for human consumption, enormous quantity of CPR is generated. This nonfood organic matter is potential good feedstock for anaerobic digestion (AD) processes that generate bioenergy.

There are numerous other reasons for the attraction of cassava crop as source of food and bioenergy:

- Cassava provides economic and subsistence value for 800–1000 million people in more than 90 countries including Angola, Barbados, Brazil, Cambodia, China, Cook Islands, Democratic Republic of Congo, Dominica, Ghana, Haiti, India, Indonesia, Lao Peoples Democratic Republic, Mozambique, Nigeria, Suriname, Thailand, Uganda, United Republic of Tanzania, and Vietnam.

- It is the fourth most important food crop in developing nations. Cassava is also world's third largest source of food carbohydrates and the top food energy supplier for tropical and subtropical regions. About 30% of all calories consumed in Mozambique come from cassava. In Zaire, cassava roots provide 60% of the daily caloric intake, while 20% of protein come from cassava leaves. In addition, Cassava can be biofortified with vitamin A, iron and zinc to eliminate hidden hunger and improve the nutritional status of vulnerable groups.

- Cassava presents numerous agro-climatic advantages and benefits as well. First, it has high biological efficiency as the edible root portion lies underground and does not require support from stems and branches. It is easily cultivated by stem cuttings for multiplication and planting purposes, and requires minimum agricultural inputs (fertilizers, pesticides, etc.). With the possible exception of sugarcane, cassava's productivity in terms of calories per unit land area per unit of time is significantly higher than that of other staple food crops;

and its production requires energy input that constitutes just 5–6% of the energy output of the entire cassava biomass.

- Cassava can be planted most time of the year and is available all year long with more than 2 years harvest window. Cassava is adaptable to various farming systems. It can be inter-cropped with beans, yams, and other annual crops. It is tolerant of various climatic conditions (e.g., high drought; temperature: 8–33 °C; rainfall: 500–6000 mm per annum; relative humidity: 15–90%; and elevation: sea level–2500 m). Cassava is also productive on soils with pH of 3–9.5. It can thus be cultivated on marginal lands where other crops such as corn, wheat, rice and sugarcane cannot be grown well. Cassava has high efficiency of photosynthetic CO_2 assimilation. The photosynthetic rate of cassava is 40–50 µmol CO_2 m^{-2} s^{-1} under high solar radiation. That of rice is around 20 µmol CO_2 m^{-2} s^{-1}.

- Cassava root is endowed with high starch content of excellent functional and structural qualities. The cassava starch can be transformed into products with huge industrial applications and is of major economic importance in Brazil, India, Indonesia, Philippines, China, Thailand, South East Asia, and in the tropical regions of the world.

- Cassava is a major ingredient for livestock feeds.

- Cassava is important in the provision of bioenergy such as bioethanol and biogas. For instance, the yield of bioethanol from cassava (6000 kg/ha) is higher than that of sugarcane (4900 kg/ha), carrot (4500 kg/ha), sweet sorghum (2800 kg/ha), Rice (2250 kg/ha), Maize (2050 kg/ha), and wheat (1560 kg/ha). A feasibility case study in Kenya using biogas engine for backup power generation showed ample savings over the use of diesel engine. Biogas engine saved 17 tons of carbon dioxide emissions, 18% reduction in net present cost, 20% reduction in levelized cost of electricity, and 30% reduction in capital cost.

Energy recycling from biomass residues and wastes is increasingly attractive because the sustainability of analyzed feedstock favors biomass waste flows over dedicatedly cultivated energy crops. Therefore, utilization of nonfood cassava processing residues such as CPR in biomethane production via the anaerobic digestion technology is prudent and beneficial. Nevertheless, in order to properly assess and quantify the value and contribution of CPR to the energy mix of cassava producing nations, establishment of Biofuel Potential (BFP) of CPR is necessary. Relatively very few studies have been published on biomethane production from cassava feedstocks. Most of the studies utilized cassava starch extraction wastewater. Other cassava feedstocks used were stillage (wastewater) from cassava ethanol production; cassava stem residue; whole cassava root; effluent from cassava flour and meal industry; and cassava peeling residue (CPR). However, CPR constitutes about 19% fresh weight of the root and is perhaps the most abundant residue from cassava root processing. It is easy to generate and does not require water usage.

Applications, Utilizations and Dividends of Biomethane from CPR

The anaerobic digestion of CPR would generate biogas, which could be used as is or upgraded to obtain more efficient biomethane. The energy content of either fuel could be put to various applications and utilities. These include:

- Fuel for stoves in cooking; boiling, frying, roasting, etc.

- Fuel for lamps in lighting; illumination, reading, playing, etc.

- Fuel for transportation; cars, trucks, sea vessels, etc.

- Electrical power in processing operations; drying, grinding, heating, pumping, refrigeration, washing, etc.

- The digester effluent (digestate) could be utilized for soil amendment and/or serve as bio-fertilizer for enhanced crop production. This was demonstrated to increase potato yield.

- Perhaps the critical humane benefits are the reduction of drudgery and burden on the one hand and the improvement of health conditions on the other hand. This is due to reduced time spent on fetching firewood and charcoal for domestic fuel, and the reduced exposure to their combustion products.

- Women and children may carry on their head 10 kg of firewood for distances up to 8 km, spending 5–6 hours per trip; 2–6 hours per day.; or 5 hours per week.

- Domestic combustion of the firewood releases health-impairing pollutants like carbon monoxide, hydrocarbons, smoke and other particulate matter. These combustion products may cause nausea, sneezing, eye and respiratory irritations; pneumonia, lung cancer, and respiratory infections; and reduced birth weight.

- Biomethane utilization reduced firewood consumption by 74% in China and 84% in Sri Lanka, thereby minimizing the drudgery, burden, and health hazards associated with use of firewood for domestic energy.

Figure presents pathways of the production and utilization of biomethane from CPR and other renewable feedstocks.

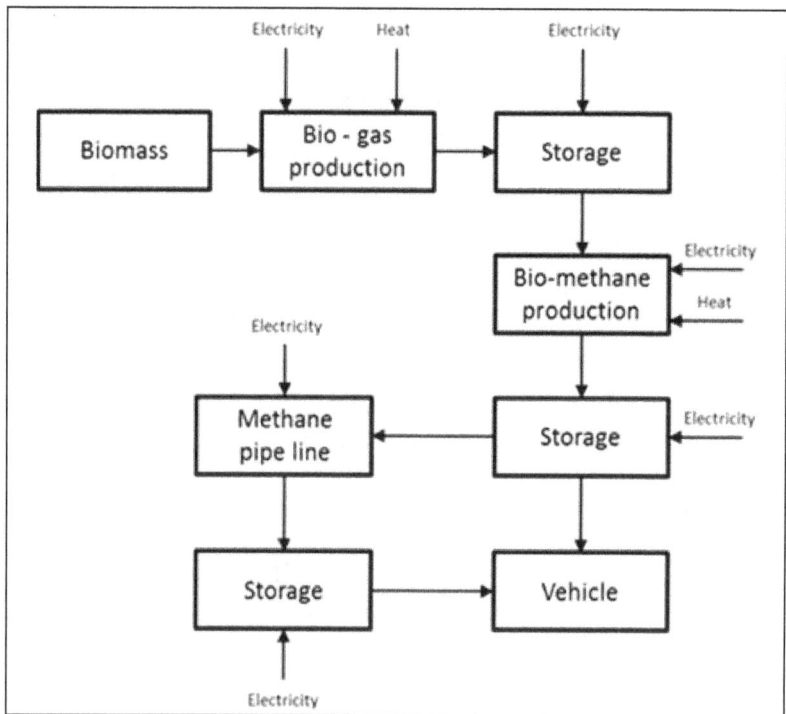

OILSEED-BASED FEEDSTOCKS

Soybean

Soybean acreage is much greater than other oilseed crops, leading to substantial soybean oil production and its availability as a biofuel feedstock.

Soybean (Glycine max) is a major crop throughout much of North America, South America, and Asia. The United States is the world's greatest producer, producing approximately 32% of the world's soybeans, followed by Brazil with 28%.

Soybeans originated in Southeast Asia, with first domestication reported in the 11th century BC in China. First planted in the United States in 1765, soybeans spread to the Corn Belt by the mid-1800s with major acreage not seen until the 1920s, when it was used mainly as a forage crop. Major U.S. expansion as an oilseed crop began in the 1940s.

Soybeans contain approximately 18% to 20% oil compared to other oilseed crops such as canola (40%) and sunflower (43%). At 48 pounds per bushel, soybean meal remains a major product from soybeans and is used for animal feed and human food.

Soybean trials.

Current Potential for use as a Biofuel

Soybean oil is currently a major feedstock for production of biodiesel (NBB). The most common method of biodiesel production is a reaction of vegetable oils or animal fats with methanol or ethanol in the presence of sodium hydroxide (which acts as a catalyst). The transesterification reaction yields methyl or ethyl esters (biodiesel) and a byproduct of glycerin.

Note that biodiesel is not straight vegetable oil burned in a diesel engine. Numerous studies between 1980 and 2000 have shown the use of straight vegetable oil, including soybean oil, causes carbon deposits and shortens engine life.

Biodiesel use in diesel engines does not have similar negative effects. Use of soybean oil for biodiesel was greatly influenced by promotion from U.S. soybean farmers through the United Soybean Board (USB) and subsequent creation of the National Biodiesel Board (NBB).

Biology and Adaptation

Soybean is a cool-season legume which can be grown from south to north throughout much of the eastern half of the United States. Soybeans and other legumes have a unique relationship with a bacteria bradyrhizobium species, which will colonize on soybean roots, forming a nodule. The two species form a symbiotic relationship in which the soybean plant provides nutrition and the bacteria fixes nitrogen from the air. This relationship reduces the need for supplemental nitrogen fertilizer in soybean production.

Soybeans flower in response to day length and temperature. Varieties grown in the United States are divided into 13 maturity groups, from maturity group 000, which is the earliest and adapted to northern regions of Minnesota and southern Canada, to maturity group X, adapted to southern regions such as south Texas. The earlier varieties bloom when days are long and nights are short, while the later-maturing varieties bloom under relatively shorter days and longer nights. Summer days are longer at northern latitudes, where early-maturing varieties will initiate flowering when days are longer. Maturity groups develop differently, and knowing the growth habit of different maturity groups can help with the crop management.

Production and Agronomic Information

Through much of the upper Midwest, soybeans are planted in April to June and harvested in September to November. Soybeans are well adapted to grow in soils similar to corn production. In many cases, soybeans are grown in rotation with corn or wheat to break insect, weed, and disease cycles.

Nutrient requirements are generally less for soybeans than other crops. Major nutrient requirements include nitrogen, phosphorous, and potassium. Much of the nitrogen is gained through a relationship with bacteria. A soil pH in the range of 5.5 to 7.0 will enhance nutrient availability and soybean growth.

Weed control is necessary to achieve optimal yields, and use of biotech seeds has eased the ability to control weeds during the growing season. Currently in the United States, over 90% of soybeans planted are herbicide resistant.

Many insects and diseases are common in soybeans grown in the upper Midwest. The most damaging pest to soybeans is soybean cyst nematode, a soil-borne parasitic roundworm that feeds on soybean roots. Insect pests include bean leaf beetle, soybean aphid, green clover worm, and spider mites. Soybean harvest begins after 95% leaf senescence when beans are at 12% to 18% moisture.

Potential Biofuel Yields

Current U.S. production of soybeans in 2009 was 3.4 billion bushels from 77.4 million acres. Average yield per acre for the United States was 44 bushels per acre (National Agricultural Statistics Service). One bushel of soybeans can yield 1.5 gallons of biodiesel (NBB). Using all U.S. soybeans for biodiesel could produce 5.1 billion gallons of biodiesel. However, using all soybean production for biodiesel has not been proposed and is not realistic.

In 2009, biodiesel production was 700 million gallons with a production capacity of 1.83 billion gallons. Based on a yield of 44 bushels per acre, an acre of soybeans could yield 66 gallons of biodiesel, compared to 69 gallons for a 1,300-lb per acre canola yield, 84 gallons for sunflower and over 600 gallons for palm oil.

Harvesting soybeans.

Production Challenges

Soybean production generally complements corn production in the upper Midwest. Both corn and soybeans enjoy a long history of production on millions of acres in the upper Midwest. This history has led to a large infrastructure of equipment, storage, rail, barge, and truck transportation.

Soybeans, like many crops, face insect and disease pests along with weather-related challenges. An emerging disease has gained much attention in recent years. Soybean rust, a fungal disease native to Asia, has spread to the soybean fields of South America and finally to U.S. soybeans. Rust control is expensive, requiring fungicide applications, and yield damage can be extreme.

Environmental and Sustainability Issues

The capacity of soybeans used for biodiesel production grew from zero to over a billion gallons per year in the past two decades. During that time, biodiesel production rose and fell depending on the price of feedstock, price of petroleum oil, and federal and state subsidies provided to the industry.

One major challenge for soybeans is the competing uses for soybean oil. Soybean oil is used in human food products, as cooking oil, and for numerous industrial applications. Soybeans account for 80% or more of the edible fats and oils consumed in the United States. Competition with other uses has caused price spikes in the soybean oil market, challenging the profitability of soybean biodiesel.

A 2009 life cycle analysis of biodiesel done by the USDA found that soy biodiesel yields 4.56 times the fossil energy needed to produce it. In comparison, petroleum diesel has a fossil energy ration of 0.84.

Rapeseed and Canola

Rapeseed is related to mustard and to other cabbage-family crops. Rapeseed has been cultivated since the 20th century B.C. Because the plant can grow with less sunlight and at lower temperatures than other crops, it was cultivated in Europe as early as the 13th century A.D.

Rapeseed oil has been used for cooking, lighting, and industrial uses. However, traditional rapeseed contains high quantities of erucic acid and glucosinolates, which make the seed meal unpalatable and possibly dangerous to livestock if fed in large quantities.

Canola — An Edible Variety of Rapeseed

Canola is an edible variety of rapeseed with a low percentage of erucic acid and low levels of glucosinolates. It was developed by Canadian plant breeders in the 1970s.

The word "canola" was coined from "Canada" and from "oleo" (oil). The term is no longer a trademark. "Canola" can be applied to varieties of rapeseed with 2% or less erucic acid and less than 30 micromoles of glucosinolates per gram of oil-free meal.

Much of the rapeseed grown in Europe is of canola quality but retains the name rapeseed probably because the word "rape" does not have the negative connotations in Europe that it does in English-speaking countries.

This small field of canola in southern Vermont is about ready to harvest.

Current Potential for use as Feedstock for Biofuel

Soybeans are the major oilseed used for biodiesel production in the United States. Edible rapeseed is the most common oilseed used for biodiesel in Europe.

Biodiesel made from canola or edible rapeseed gels at a lower temperature than biodiesel produced from other feedstocks, making canola biodiesel a more suitable fuel for colder regions. University of Idaho research showed that canola biodiesel had a "cloud point" of 1 °C and a "pour point" of -9 °C.

The cloud point is the temperature of the fuel at which small, solid crystals can be observed as the fuel cools. These crystals will clog vehicle filters. The pour point refers to the lowest temperature at which there is movement of the fuel when the container is tipped. Because canola biodiesel has a slightly lower cloud point and pour point than soy biodiesel, and a much lower cloud point and pour point than biodiesel made from animal fats, canola biodiesel is useful in cold climates.

Canola and rapeseed contain about 40% oil and have a high yield of oil per acre: 127 to 160 gallons per acre, compared to 48 gallons per acre for soybeans.

Canola oil is high in oleic acid, which makes it competitive with other cooking oils, a market in which it is well established. The oil is also a high-grade lubricant and fuel additive; conversion to biodiesel, therefore, is just one of its several potential end uses.

Canola meal (what's left after the oil is extracted) is a good source of protein, containing 38 to 42% protein and a favorable balance of amino acids. It can be used as a feed additive for livestock rations.

Industrial rapeseed makes a biodiesel with very good low temperature performance. University of Idaho research showed that rapeseed biodiesel had a cloud point of 0 °C and a pour point of -15 °C. However, comparatively little of this crop is grown because the market for canola and edible rapeseed is much larger than the market for industrial rapeseed.

Industrial rapeseed contains more long-chain fatty acids than canola. Therefore, sometimes industrial rapeseed biodiesel turns out to be slightly more viscous (thicker). In this case, the rapeseed biodiesel can be blended with other fuels (such as canola or soy biodiesel) in order to meet the specification.

Canola meal.

Biology and Adaptation

Rapeseed and canola divide into two main species: Brassica rapa, known as "Polish type," and Brassica napus, known as "Argentine." Canadian breeders have also developed a low erucic acid, low glucosinolates variety of brown mustard (Brassica juncea).

There are both spring and winter (fall planting) types in canola and rapeseed species. The species differ in agronomic characteristics and yield. These differences must be evaluated when selecting a variety to grow.

Production

Canola in Franklin County.

In temperate climates such as the Pacific Northwest, canola/rapeseed can be planted either in the fall or spring. Fall-planted canola or rapeseed can develop more extensive root systems and is more drought hardy, but excessively cold winter weather or wet winter growing conditions can reduce yield potential, so the advantage may lie with spring planting dates. Canola must be planted in time to ensure maturity before the onset of hot weather. Winter canola must be planted in time to ensure significant plant development (six leaves or more) before hard freezing weather.

Most canola in the United States is produced in North Dakota. Responses to fertilizer and soil fertility are similar to those for small grains; however, canola is a heavy user of sulfur. In a 2,000 lb/acre crop, for example, about 12 and 15 lb/acre of sulfur are in the straw and seed, respectively. Canola competes well with weeds, and herbicides are registered for use in the crop.

Seed size ranges from 80,000 to 135,000 seeds/lb, depending on variety. Canola is handled and stored like flax; tight containers are necessary to avoid loss in transit.

Potential Yields

Yields of oil per acre vary from about 75 gallons per acre to about 240 gallons per acre.

In Oregon canola trials, yields ranged from 1,900 to 4,800 pounds of seed per acre. Since canola is about 40% oil, and since a gallon of vegetable oil weighs about 8 pounds, this comes out to about 95 to 240 gallons of oil per acre.

The 2009 canola trials in North Dakota resulted in an average yield of 1,900 pounds of seed per acre, with an average oil content of 45%. This works out to about 107 gallons of oil per acre.

Recent trials in Maine (where the crop is relatively new) resulted in 75 to 100 gallons of oil per acre.

Trials in Minnesota resulted in an average of about 96 gallons of oil per acre. The average percentage of oil in the seeds was 46%.

Production Challenges

Since it is a Brassica crop, canola can cross pollinate with other Brassicas such as rutabaga, Chinese cabbage, broccoli rabe, and turnip unless buffer distances are adequate. In addition, it is problematic to grow canola among infestations of mustard-family weeds.

Canola grows on most soil types but requires good drainage. The emerging crop is very susceptible to soil crusting; seedbed preparation is important. Canola is susceptible to blackleg and Sclerotinia stem rot. If not rotated with resistant crops, seed treatment may be necessary.

Seed shattering at harvest is a potential problem, so crops commonly are swathed or "pushed" (mechanically bent over without cutting the stem) when seed moisture is about 35%.

Palm Oil

Palm oil together with corn, rapeseed, soybean and sugar cane are viable feedstocks for use as first generation biofuel.

According to the Food and Agriculture Authority (FAO) from a sustainability perspective, bio-fuels offer both advantages (energy security, GHG reductions, reduced air pollution) and risks (intensive use of resources, monocultures, reduced biodiversity and even higher GHG through land use change). Therefore, to measure biofuel's sustainability, economic, environment and so-cial sustainability factors must be considered.

In terms of yield productivity, sugar cane and palm oil rank the highest. Sugar cane yields 6,000 litres of biofuel per hectare (l/ha), followed by oil palm and sugar beet (5,000-6,000 l/ha) but palm oil is superior as it has 27% higher energy content (30.53 MJ/l) than ethanol from sugarcane (24MJ/l). Moderately efficient feedstock's such as corn, cassava and sweet sorghum yield 1,500-4,000 litres of biofuel per hectare(l/ha). Rapeseed, wheat and soya are the least efficient, yielding less than 1,500 l/ha. Interestingly, it is these moderate to low efficient feed-stocks that are used in countries with mandated biofuel programmes; in the US biofuels from soya and corn are used while in EU rapeseed is the preferred choice. Although the use of these feedstocks may not be economical, they become viable due to subsidies and mandates set by the governments.

FAO's search found sweet sorghum as another possible alternative biofuel feedstock. Although it can rival sugar cane in terms of productivity, it requires quick processing after harvesting and poses challenges for transportation and storage given the bulkiness of the crop.

Jatropha was thought to be a plausible biofuel that would put to rest the "food versus biofuel" debate. As the first generation biofuels are also food crops, there was a fear that using them for biofuel would create a shortage in the food supply and drive up food prices. According to FAO jatropha would require intensive crop management to be successful which, in turn, would result in competition for top farm land. In reality, any crop grown as a source for biofuel feedstock will still compete with food crops for land and water resources. In the end, economics will trump agronomy in making the choice.

In countries where cassava is grown widely, it is a staple food crop. In these countries, the poten-tial to develop it into biofuel is impeded by limited processing technologies and underdeveloped marketing channels. It is unlikely that it will become a large scale biofuel source.

With regard to advanced biofuels (including cellulosic ethanol), it has not reached the stage to be viably produced commercially. Dedicated energy crops (e.g. alfalfa, swithgrass, miscanthus), fast-growing short rotation trees (e.g. poplar, willows, eucalyptus) and wood and agricultural resi-dues offer great potential. Currently, economics and high capital investment for new supply chains remain serious obstacles for second generation biofuels. It is also cautioned that the advent of second generation biofuels would create pressure for land to produce such crops and worsen the competition with food crops.

Economic Sustainability

Economic sustainability requires long-term profitability, minimal competition with food produc-tion and competitiveness with fossil fuels. As biofuel programmes are supported by subsidies and mandates, these factors mask the true economic assessment. It is, thus, difficult to assess the long run economic viability of biofuel systems. Nevertheless and FAO opines that despite the added certification cost, feedstock for biofuels made from palm oil and sugar cane produced by developing countries are still able to compete in the European market. This is a clear indication of the economic viability of these two prime biofuel feedstocks.

Environment Sustainability

The issues tied up with environment sustainability may be global (e.g. climate change, GHG mitigation, renewable energy,) and local (e.g. water pollution, soil quality, erosion, air pollution). Life cycle assessment methods are often used to study these aspects but the methodologies are not standardized and cannot adequately quantify indirect land use changes.

Fossil energy balance, which is the ratio between renewable energy output and fossil energy input is a good factor to compare biofuel sources. Topping the list is palm oil biodiesel with a fossil energy balance of 9.0. This means that a litre of palm oil biofuel contains 9 times the amount of energy as was required for its production. Sugar cane has values ranging from 2.0 to 8.0. Other feedstock's; rapeseed, soya and corn have values which fall within 1 to 4.

A major portion of the high fossil fuel energy input to produce temperate biofuels is that they require large quantities of fertilizers; thus, the fear of endangering environment sustainability, e.g. water pollution, at the local level. In comparison with soya and rapeseed, oil palm requires lower inputs of fertilizers and agrochemicals.

Sugar cane has the lowest water footprint, with an average of 29 m3/GJ. while oil palm (75 m3/GJ), sunflower (72 m3/GJ) and soya (99 m3/GJ) have medium water footprints. Rapeseed has a very high water footprint (average 131 m3/GJ).

Irrespective of which biofuel feedstock is grown, there is concern that biomass (for conversion into biofuels) production under intensive agriculture can have negative impacts on biodiversity, including habitat loss, expansion of invasive species and contamination from fertilizers and herbicides, especially if they are monoculture systems. According to FAO, cultivation of biofuel production systems will destabilize the original biodiversity composition. For oil palm, there is the concern that if large areas of planting in the future are carried out on peat or tropical forest, the carbon debt will be high. The solution as practised in Malaysia is to commit a minimum of 50% of the total land area to be out of bounds for agriculture and maintained as permanent forest to sustain the mega-biodiversity status of the country.

Social Sustainability

The social dimension of biofuel sustainability relates to the potential for rural development, poverty reduction and inclusive growth. The Social Impact Assessment should be used as a tool to measure social sustainability. The FAO report did not compare the various kinds of biofuels in this aspect. This lies in the difficulty of translating social sustainability standards and criteria into measurable indicators. As such, most present systems of measuring social sustainability only pay attention to social aspects which have negative impacts; such as child labour, minimum wages or calling for adherence to national laws or international conventions.

FAO states that critical factors e.g. health implications, poverty eradication or smallholder inclusiveness are not included. Social sustainability must move away from just focusing on a few negative impacts and include these factors and development goals where local communities share sustainably in the economic benefits derived from biofuels in comparison with other alternatives.

Jatropha

Considering environmental issues and to reduce dependency on fossil fuel many countries have politicized to replenish fossil fuel demand from renewable sources. Citing the potential of Jatropha mostly without any scientific and technological backup, it is believed to be one of the most suitable biofuel candidates. Huge grants were released by many projects for huge plantation of Jatropha (millions of hectares). Unfortunately, there has been no significant progress, and Jatropha did not contribute much in the energy scenario. Unavailability of high-yielding cultivar, large-scale plantation without the evaluation of the planting materials, knowledge gap and basic research gap seem to be the main reasons for failure. Thus, the production of Jatropha as a biofuel has been confronted with various challenges such as production, oil extraction, conversion and also its use as a sustainable biofuel.

Jatropha belongs to the family Euphorbiaceae and has 175 species. It has originated from tropical America and has spread all over the tropics and subtropics of Asia and Africa. Throughout the world, more than 1,000,000 ha of Jatropha have been propagated. Majority (85%) of them are in the Asian countries, i.e., India, China and Myanmar; the remaining, 12% in Africa and 2% in Latin America (Brazil and Mexico). India is the largest cultivator of Jatropha. In the ancient times, Jatropha has been used in various fields, such as storm protection, soil erosion control, firewood, hedges and traditional medicines. The seed oil of Jatropha is also used as lamp fuel, soap manufacturing ingredient, paints and as a lubricant. The characteristics of Jatropha seed oil match with characteristics of diesel, thus it is called a biodiesel plant. Jatropha grows on diverse wasteland without any agricultural impute (irrigation and fertilization) and has 40−60% oil content. Easy propagation, rapid growth, drought tolerance, pest resistance, higher oil content than other oil crops, adaptation to a wide range of environmental conditions, small gestation period, and optimum plant size and architecture (which make the seed collection more convenient; actually inconvenient) are some characteristics of Jatropha, which makes it a promising crop for biofuel. Although Jatropha ranked behind palm (palm > Calophylum inophyllum> Cocus sp. > Jatropha) according to annual oil yield/hectare, it is favoured as a non-edible feed stock. A number of earlier reports, proceedings, expectations and assumptions predicted that the seed yield of Jatropha range from 2 to 5 Mg/ha and even 7.8 to 12 Mg/ha- without any scientific and technological backup.

There is a complete mismatch between theoretical expectation and actual seed production of Jatropha in field conditions. The research on Jatropha opened the floodgate to the scientific community to grab funds and publish papers in high impact journals because seed oil of Jatropha has characteristics of biodiesel and this crop was non-native of arid, semiarid and subtropical regions. Singh and co-authors depict the expectation and contribution picture from Jatropha policy. The reported yields of Jatropha in field conditions in India, Belgium, South Africa and Tanzania, are 0.5−1.4 mg/ha/yr, 0.5 mg/ha/yr, 0.35 mg/ha/yr and 2 mg/ha/yr, respectively. The less productivity is because of unavailability of suitable high yielding varieties, large-scale plantation without evaluating the genetic potential of planted materials, consideration of Jatropha as no/low impute crop and lack of knowledge on agronomy.

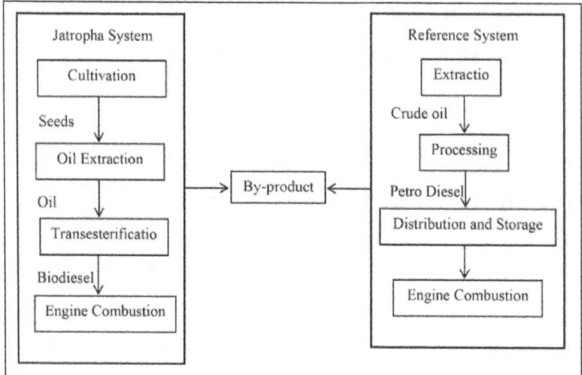

Potentials of Jatropha

Jatropha has multiple uses. Jatropha seed oil possesses biodiesel and jet fuel production potentials. Its wood, leaves and fruits have been using as firewood in rural areas. It also has industrial applications. Preparation of soap and cosmetics, and dyeing clothes and fishing nets are some of its common applications. Traditionally, Jatropha has been known as a medicinal plant. The therapeutic compounds from Jatropha can be used as anti-microbial, anti-inflammatory, healing, homeostatic, anti-cholinesterase, anti-diarrheal, anti-hypertensive and anti-cancer agents in modern pharmaceutical industry. As it contains toxins, before using Jatropha and/or its derivatives as a therapeutic agent, toxicological studies must be conducted. Jatropha seed cake can supplement animal feed and organic fertilisers as it bears higher percentage of protein and other nutrients. Soil erosion control and used as hedges are prehistoric uses of Jatropha.

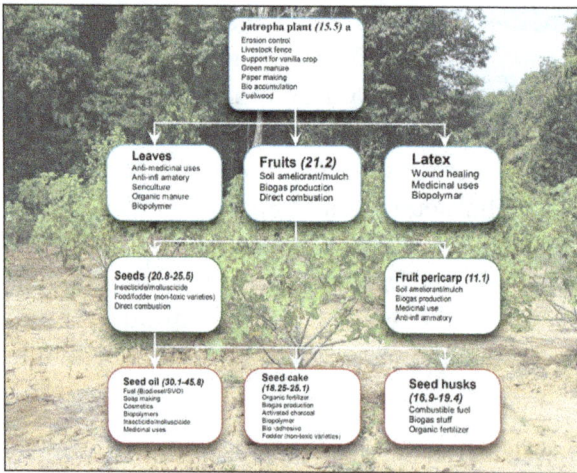

Multipurpose uses of Jatropha curcas.

Jatropha as Energy Source

Various features like, ease of production, sustainability and environmentally friendly nature of biomass draw attention as a potential renewable energy to replenish fossil fuel demand. Among the crops identified as energy crops for first generation biofuels, Jatropha curcas L. (JCL) has been acknowledged as one of the promising candidates.

Parts of Jatropha plant, like wood, fruit shells, seed husks and kernels, are used to produce energy. Raw oil is the major resource obtained from Jatropha. Depending on the variety/cultivars, decorticated seeds contain 40–60% oil. The oil is utilised for many purposes, such as lighting, lubricating, making soap and most importantly as biodiesel. Biodiesel from Jatropha comply with European biodiesel standards.

Table: Chemical and physical properties of Jatropha oil.

Parameter	Jatropha oil
Density at 15 °C	0.920 g/cm³
Viscosity at 30 °C	52 cSt
Flash point	240 °C
Fire point	274 ± 3 °C
Cloud point	9 ± 1 °C
Pour point	4 ± 1 °C
Cetane number	38
Caloric value	38.20 MJ/kg
Conradson carbon residue	0.8 ± 0.1 (%w/w)
Hydrogen	10.52 (%w/w)
Sulphur	0 (%w/w)
Oxygen	11.06 (%w/w)
Nitrogen	0
Carbon	76.11 (%w/w)
Ash content	0.03 (%w/w)
Neutralization number	0.92 mg KOH/g
Saponification value	198
Iodine number	94
Monoglycerides	Not detected
Diglycerides	2.7% m/m
Triglycerides	97.3% m/m
Water	0.07% m/m
Phosphorus	290 mg/kg
Calcium	56 mg/kg
Magnesium	103 mg/kg
Iron	2.4 mg/kg

There are approximately 24.60%, 47.25% and 5.54% of crude protein, crude fat and moisture, respectively, in Jatropha oil. Both saturated and unsaturated fatty acids are present in the oil. The major saturated fatty acids are Palmitic acid (16:0) at 14.1% and stearic acid (18:0) at 6.7%, oleic acid (18:1) at 47.0% and linoleic acid (18:2) at 31.6%. The usefulness of Jatropha oil and its esters instead of petro-diesel has been reported. The energy value of Jatropha seed oil (39MJ kg-1) is higher than anthracite coal and is comparable to crude oil.

By mass, the shells bears about 35–40% of the dry fruit, that is, 60–65% of the seed weight. There are approximately 42% husk and 58% kernel in a seed. The gross energy value of Jatropha seed is 24 MJ kg–1 which is higher than lignite coal and cattle manure and is comparable to corn cobs.

The first step of oil extraction is the mechanical removal of the shell from the fruit. Jatropha seed shell contains cellulose (34%), hemicelluloses (10%) and lignin (12%). Approximately 11.1 MJ energy is driven from one (1) kg seed shell of Jatropha.Ash (4%), volatile matter (71%) and fixed carbon (25%) are the components of seed husk. Approximately, 16 MJ energy is driven from one (1) kg seed husk, which is comparable to wood.

Less energy expenditures and the prospect of using a cheap substrate make hydrogen (H_2) gas a lucrative source of future renewable energy. Lignocellulose biohydrogen can be produced by the fermentation of de-oiled Jatropha solid waste (DJSW) and Jatropha seed cake that contains lignocellulose. Kuma and co-workers reported highest achievable cumulative hydrogen production (CHP) of 296 mL H_2 by the fermentation of de-oiled Jatropha waste under optimum conditions. The reported optimum conditions are; substrate concentration 211 g/L, pH 6.5 and temperature 55.4 °C. Lopes and co-workers produced 68.2 mL H_2/gVSiJSC biohydrogen by dark fermentation of seed cake by a pure strain of the bacteria Enterobacter aerogenes without pretreatment of the substrate. In the viewpoint of energy saving, it is significant.

Limitation of Jatropha as a Biofuel Crop

- A good commercial variety with a higher yield and disease resistance is still lacking.

- High fluctuation of yield among trees.

- It requires proper irrigation and nutrients for fruiting, though it can survive on insufficient irrigation and nutrients.

- Relatively long gestation period: it requires 3–5 years to become commercially productive.

- The presence of toxic components limits its use as feed and therapeutic agents.

- Recent study reveals that Jatropha is susceptible to pests and diseases.

- Jatropha is sensitive to frost and water logging.

- Jatropha may be host for some diseases (cassava diseases).

- High viscosity of Jatropha seed oil limits its use in cool climate conditions.

- In certain environments, Jatropha may create weed problem.

Jatropha Production Challenges

Poor Seed Yield

Related experts suggest that the Jatropha seed yield of 4–5 Mg/ha/yr is needed for the commercial viability of the industry. If the usual seed yield of 3.75 Mg/ha with 30–35% oil content or 1.2 Mg/ha oil yield only then Jatropha would compete with soybeans (USA 0.38 Mg oil/ha) and rapeseed (Europe 1.0 Mg oil/ha). However, there is high flocculation of the unit seed yield and seed oil content of Jatropha. A number of authors reported that the low seed yield and the low seed oil content are the one of the most important barrier for commercial viability of Jatropha biodiesel industry. In India, a different location trial at diverse agro-climatic regions was conducted and the average seed yield was recorded as 0.5–1.4 Mg/ha/yr after 5 years of plantation. A similar result was observed from plantation of 24 elite accessions with good plant architecture (height and branching pattern) in sodic soil. In Belgium, the average seed yield was reported as 0.5 Mg seed/ha after 4 years of plantation, using the best known production techniques. Recent assessment revealed that globally the average seed productivity of Jatropha is 1.6 Mg/ha which is equivalent to 0.475 Mg/ha/yr biodiesel productivity, which is not a safe position for the industry to be economically feasible. In South Africa, the highest seed yield was 0.35 Mg/ha after 5 years of plant growth. A Jatropha silvi-pastoral production system in central-west Brazil where hybrid seeds were used, however, it could not ensure any significant seed yield, against the expectation of 2.4 kg/plant. In Tanzania, a negligible gain at US\$ 9 ha−1 with yields of 3 Mg/ha and a loss of US\$ 65 ha−1 on lands with yields of 2 Mg/ha of seeds after 5-year investment were obtained. In Panzhihua, China, Jatropha could not change local energy scenario and the industry has been confronted by a number of risk factors.

Developing of a higher yielding and more oil containing variety is one of the main effective solutions. However, a good commercial variety is still missing. variety breeding is one of the main hurdles for Jatropha planting.

Actually, the current Jatropha breeding program is limited to conventional breeding and surveying of germplasm resources of wild Jatropha plants. However, the study of modern biotechnology application on Jatropha improvement is limited. Particularly, studies on cloning, expression and biological function annotation for Jatropha genes, which are responsible for economical traits, are largely absent.

The enhancement of unit seed yield of Jatropha for commercial use should be the main objective of cultivation. Therefore, the techniques of Jatropha cultivation refers to many field practices such as propagation, site preparation, tree density and canopy control, insects and diseases control, fertilization and irrigation management, cropping treatments. Few studies on planting techniques and poor management for planting base limit large-scale plantation of Jatropha. However, there is limited research to demonstrate precisely and scientifically the impact of field operation on the seed yield of Jatropha. Moreover, there are no/a few detailed reports on field observation on the seed yield under different treatments of cultivation techniques. For example, data on tree density for Jatropha cultivation, canopy pruning intensity and frequency, insecticide effect as well as fertilization and irrigation efficiency are largely absent in the literature.

Consider as Low Impute Crop

J. curcas is believed as a low input crop because of its ability to grow on barren land. However, it needs adequate nutrients as fertilizer and rainfall or irrigation for growing as a productive crop. On the other hand, excessive fertilization and irrigation may cause vegetative growth (biomass production) at the cost of fruit production. Moisture and nutrients have larger influence on the seed yield and oil productivity from the plantation on marginal lands. The plant growth and the seed yield of J. curcas were significantly increased under irrigated conditions as compared to non-irrigated conditions. It was observed that there was 750 kg/ha yield under irrigated conditions at the same time only 450 kg/ ha was recorded under rainfed conditions from 3-year-old plantations. Application of nitrogen and phosphorus increased the growth, seed yield and oil yield of J. curcas. Furthermore, another report by BAIF Development Research Foundation showed that there was about 500 kg/ha seed yield under rainfed conditions in the fifth year of plantation. However, after regular irrigation of the same plantation, in next year the seed yield was recorded about 1200 kg/ha.

The systematic studies for yield improvement, the agronomy (especially the irrigation and nutritional requirements) in different agro climatic conditions have not been adequately addressed, despite advocacy for large-scale plantation of J. curcas.

Pest and Disease Susceptibility

Control of insects and diseases is particularly one of the most important technical issues which could seriously shape Jatropha cultivation. Though it was claimed that Jatropha is free of pests and diseases, the current study do not support the claim. Recent studies reported that the plants were susceptible to viral infection (Cucumber mosaic virus), insect attack, rodents, powdery mildew, leaf spots, insect defoliations and fungal diseases of the soil. In Belgium, leaf miner Stomphastis thraustica, the leaf and stem miner Pempelia morosalis and the shield-backed bug Calidea pana-ethiopica are the major pests affecting Jatropha. Fruit sap sucking predators Scutellera perplexa and Maconellicoccus hirsutus have recently been investigated in India. These infections caused approximately 60–80% damage to the standing Jatropha crop at different study sites.

The photographs (A and B) show viral incidence in Jatropha.

Moreover, monocropping could result in the spreading of insects and diseases. In Panzhihua, China Yu and co-workers found 24 species of insects and diseases that are affecting Jatropha. Wu and co-authors reported eight diseases and seven species of insects on Jatropha in the dry-hot valley of Yunnan Province. Jatropha monoculture expansion may spread insects and diseases.

Jatropha Breeding Objectives

- Dry matter (increased fruit and carnal size) accumulation in fruits rather in cost of vegetative growth.

- Higher female flowers ratio per inflorescence for more fruits.

- Flowering and fruit maturity synchronizing for mechanisation of harvesting.

- Bigger seeds and more oil contents.

- More branching to produce more flowers, fruits and seeds.

- Oil quality improvement.

- Development of non-toxic variety for safe use.

- Disease and stress tolerant cultivar development.

- Improve plant architecture for deeper and smaller rooting.

Oil Extraction Challenges

Jatropha seeds contain 40–60% of oil depending on the variety. The first step of oil extraction is the removal of shells from the seeds after collecting the ripe fruits from trees. Seed oil can be extracted manually, mechanically, chemically and enzymatically. The oil extraction process is shown in figure. Oil can be extracted by mechanical pressure, solvent extraction and enzymatic degradation of kernel. Mechanical extraction yields about 90% of total oil from the seed. Solvent and enzymatic extraction yield almost 100% of oil from the seed. However, these are complex processes and take long time. Solvent extraction involves handling of large volume dangerous chemicals. Commercially suitable enzyme(s) is still not available for enzymatic extraction of oil from seed kernel till date.

In the mechanical process, a machine is used to exert pressure on seeds for the removal of oil. After cleaning and checking, the seeds are fed into the hopper of the machine. For Jatropha seed 0.41 L of oil is extracted from 1 kg. Mechanical parameters and pretreatment of seeds affect oil yields. The effects of treatment and physical parameters on the oil extraction are shown on figure. The amount of oil that can be recovered from the seeds is affected by:

- Throughput: It is the amount of seed crashed per hour (kg/h). The higher throughput recovers less amount of oil per kg of seeds, because of short time exposure of seeds to pressure. It can be regulated by altering the turning pace of the screw throughput.

- Oil point pressure: It is the amount of pressure necessary to start oil flow from the seeds. If it is possible to reduce the oil point pressure, the oil extraction becomes easier.

- Pressure: The more the pressure, the more the oil recovered from the seeds. But oil recovery at high pressure brought more solid particles with oil. It makes the removal of solid particles more difficult. A pressure range of 50–150 bar is considered as the optimum operating pressure for engine-driven oil extraction.

- Nozzle size: A smaller pore causes higher pressure and therefore a higher oil yield. An ideal nozzle size is needed for smooth operation.

- Hull content of the seeds: Less energy should be used for pressing seeds so that there are no hull fibres in the crude oil. However, it appears like paste inside standard expellers, which sticks to the worm and keeps rotating along with it.

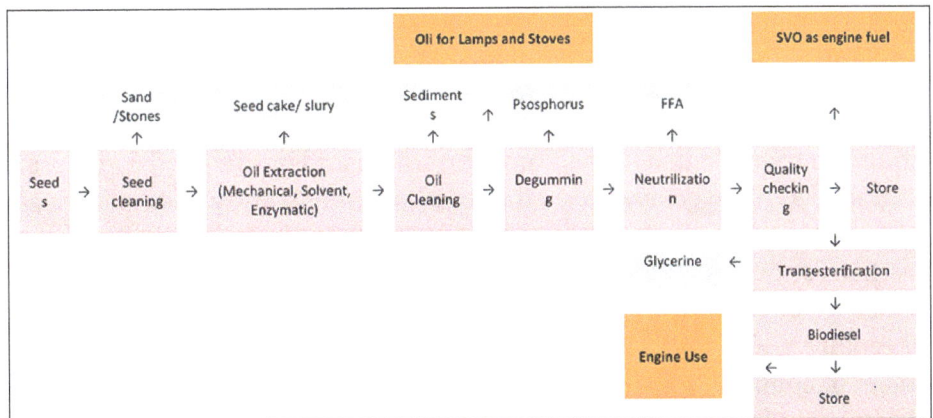

Oil extraction steps and use.

Pre-treatments		Oil Yield	Pressure	Temperature	Throughput	Energy/litter
heating		▲	▼	▲		▼
boiling		▼	▲	▲		▲
flaking		▼	▲	▼		▼
moisture content	▼	▲	▲	▲	▼	▲
hull fraction	▼	▼	▼	▼	▼	▼
Mechanical factors						
RPM	▼	▲	▲		▼	▲
restriction size	▼	▲	▲	▲		▲

Pre-treatment and mechanical factors effect on seed oil recovery.

The mechanical method is easier and less expensive but produces less oil (8–9%). Heat is generated during the process that affects the quality of biodiesel. A high efficient oil recovery (90–98%) technique, solvent extraction, is the most widely used. However, high energy input and toxicity of solvent used are major disadvantage of this technique. Enzyme-based techniques may be the solution. For extraction of oil from Jatropha seeds, aqueous enzymatic oil extraction (AEOE) is a promising technique. Plant cell walls are composed of a complex chemical structure. Enzymes that present in the system break cell walls and oil bodies and accelerate oil recovery. This eco-friendly process does not produce volatile organic compounds as atmospheric pollutants. Prolonged reaction time is the major disadvantage associated with AEOE. Moreover, suitable commercial enzyme is not available till date.

FIBER AND GRASS CELLULOSIC FEEDSTOCKS

Miscanthus Giganteus

Giant miscanthus is a perennial, warm-season Asian grass with the C4 photosynthetic pathway. Miscanthus species have been used for forage and thatching in Japan for thousands of years, managed through burning and grazing in vast prairies similar to those managed by Native American tribes in the central United States. Giant miscanthus was first collected in the 1930s as a horticultural specimen and is still planted in gardens because of its straight, tall stems and striking silver flowers. In the search for ideal bioenergy crops following the oil crisis of the 1970s, evaluations to determine the biomass yield potential of giant miscanthus began across Europe.

Current and Potential use as a Biofuel

Giant miscanthus has been studied in the European Union and is now used commercially there for bedding, heat, and electricity generation. Most production currently occurs in England but also in Spain, Italy, Hungary, France, and Germany. Recently, Japan and China have taken renewed interest in this native species and started multiple research and commercialization projects. In the United States, research began at the University of Illinois at Urbana-Champaign in 2001 and has expanded rapidly to other U.S. universities. Giant miscanthus has been proposed for use in the United States in combined heat and power generation, as a supplement or on its own. It is also a leading candidate feedstock for cellulosic ethanol. Although it is widely touted for cellulosic ethanol, giant miscanthus has traits that likely make it better suited for thermochemical conversion processes over biological fermentation, at least under existing technology.

Table: Traits of giant miscanthus that make it more (+) or less (-) amenable to conversion to biofuel.

Trait	Biological Conversion	Thermochemical Conversion
Low moisture at harvest (10-25%)	+/-	+
Low free sugar content	–	+
Low nitrogen content	–	+
High lignin content	–	+

The main feature distinguishing giant miscanthus from other biomass crops is its high lignocellulose yields. In the United States, giant miscanthus can yield more annual biomass than any other major biomass crop save Saccharum spp. (sugarcane, energycane) and has a much broader growing range. In small trials in Illinois, giant miscanthus yielded more than in European trials and two to four times more than native switchgrass . At average yields seen in Illinois trials, giant miscanthus has the potential to supply all the advanced biofuel required under the Energy Independence and Security Act using only the same land area currently devoted to producing corn grain ethanol. This means that giant miscanthus could meet biofuel goals without bringing new land into production or displacing food supply.

Table: Biomass production, potential ethanol production, and land area needed for different potential bioenergy systems to reach the 35 billion gallon U.S. renewable fuel goal.

Feedstock	Harvestable Biomass (Tons/acre)	Ethanol (gal/acre)	Million Acres Needed for 35 Billion Gallons of Ethanol	% 2006 Harvested U.S. Cropland
Corn grain	4.5	456	12.6	24.4
Corn stover	3.3	300	19.1	37.2
Corn total	7.8	756	7.6	14.8
Switchgrass	4.6	421	13.6	26.5
Miscanthus	13.2	1198	4.8	9.3

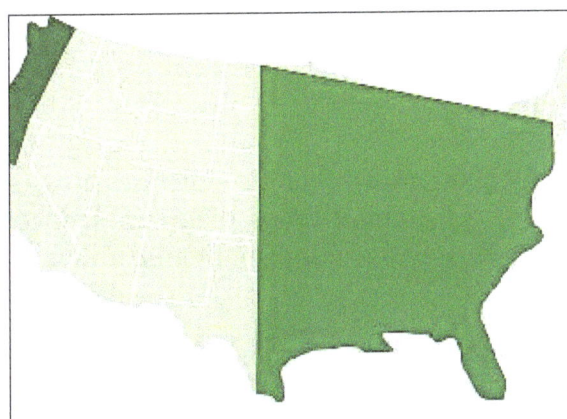

Approximate growing range of giant miscanthus in the United States.

Giant miscanthus is best suited for areas with at least 30 inches of rain per year, with better results as rainfall increases. Giant miscanthus can tolerate wet soils, and dry matter yield is directly correlated to seasonal precipitation.

Biology and Adaptation

Giant miscanthus is a cross between M. sacchariflorus, a tetraploid species, and M. sinensis, a diploid species. When crossed, they create the sterile triploid hybrid M. x giganteus. M. sacchariflorus is characterized by fast-growing rhizomes and high productivity in warm, wet areas. M. sinensis relies primarily on seed for reproduction and is found in montane environments that frequently have cold winters. The progeny of this cross, the hybrid giant miscanthus, can be thought of as "the mule of the plant world" – that is, it is sterile and bigger than either parent. Further, it inherited good cold tolerance and is currently the most productive crop known for cool, temperate regions of the world. Giant miscanthus is a close relative of sugarcane but has low concentrations of sucrose, and Miscanthus spp. have been used to breed disease resistance and cold tolerance into sugarcane varieties. Moreover, there is a high degree of genetic similarity between the genera, suggesting that genomic advances in sugarcane could also be used to improve miscanthus.

Hybrid origins of giant miscanthus

Production and Agronomic Information

Field Preparation

Best sited on well-drained soils, giant miscanthus also tolerates heavy soils and periodic flooding. When grown on marginal sites, as with most crops, yields are reduced but still considered high compared to other perennial grasses. Best yields of giant miscanthus are likely to occur on ground suitable for annual row crops.

A fall cover crop of small grains may be used to protect the soil over winter and provide weed suppression in the spring. The cover crop should be killed prior to planting giant miscanthus in the spring. A burn-down herbicide application may also be necessary to control spring weeds before planting. The whole field may be tilled prior to planting or, to better protect the soil and prevent weeds, tilled only in strips wide enough for planting (this will depend on planting equipment). Soil should be finely tilled to a depth of at least 6 inches.

Planting

Planting technology is one of the major limitations to giant miscanthus use in the United States today for two reasons: limited plant material and limited planting equipment.

First, finding quality plant material of known genetic background is difficult. There are only a few purveyors of giant miscanthus in the United States today. Verifying that the material sold is actually giant miscanthus is essential to ensure that the sterile triploid hybrid is used and not a fertile variety that could become an invasive liability. Second, dedicated machinery for planting giant miscanthus is being developed but is not currently available in the United States. Giant miscanthus can be planted from rhizomes dug straight from a mother field or from greenhouse-grown plants, called plugs. Certain types of vegetable transplanters are appropriate for both rhizomes and plugs, but there are issues to consider with each. For example, the rhizomes must be cleaned and sized to fit through the transplanter before planting. If planting plugs, it is critical to apply water at planting to ensure good survival. Rhizomes should be planted 2 to 4 inches deep and well covered. Plugs should be planted with the root ball below the soil surface. Both can be planted anytime after the frost-free date, typically by May 1 in the midwestern United States.

We recommend planting at conventional spacing for available equipment, with good success using 30 inches between and within the rows. It appears that equal spacing around the plant gives better growth than planting at a higher rate within the row as is done with many annual crops. This spacing will also make it easier to include field cultivation as a weed control option.

Pest Control

Controlling weeds in new plantings of giant miscanthus is necessary to develop a quality establishment. In research trials, good weed prevention has been realized through preplant and preemerge applications of pendimethalin and atrazine, with reapplication as needed to prevent growth of grasses and small-seeded broadleaves. Typically, little to no herbicide is needed by the third year after planting. It is essential that perennial grass weeds not become established in a newly planted field. Quinclorac and alachlor have also been used to control postemergent grass weeds with some success. Broadleaf weeds are easily controlled with 2,4-D. Because giant miscanthus is a new crop and on very limited acreage in the United States, currently no herbicides are specifically labeled for use. Please refer to product labels for complete legal information. At present, there are no commercial pests of giant miscanthus in the United States or Europe. Given the clonal nature of the crop, any pest issues that do arise may become serious.

Fertility

The importance of fertilizer to increasing harvestable yield is still not clear. Although productivity is often higher on more fertile soils, frequently higher yields are realized on poorer soils if other environmental conditions, such as temperature, are favorable. Numerous studies have investigated yield responses to nitrogen (N) fertilizer with varying results, frequently showing no significant yield increase, even after several years of biomass removal. Yield increases have been most commonly seen on sandy soils under irrigation, for example, with responses less clear under water limitation. In Ireland, a stand growing on marginal land for 16 years showed a response to potassium fertilization but not to N suggesting other nutrients may limit yield before nitrogen. The apparent N use efficiency of giant miscanthus is thought to result from effective internal cycling of the N, i.e., N taken up by the crop during active growth is translocated to the rhizomes during senescence, where it is stored and then used again during the following year's growth.

Table: Preliminary fertility requirements for giant miscanthus in the Midwest.

Parameter	Recommended Range
Soil pH	6-8
N	5.5-9 lb/ton removed
P	1.5 lb/ton removed
K	5.5-9 lb/ton removed

The wide variety of results from the limited number of studies done on giant miscanthus fertility have made best management recommendations difficult, with most authorities advising a regime modified from forage management.

Harvest

A variety of conventional hay forage equipment
is suitable for harvesting giant miscanthus.

Giant miscanthus can be harvested with a variety of conventional hay or silage equipment. To fully realize nutrient cycling during senescence, the crop should be allowed to fully dry down before harvest. The typical harvest window for giant miscanthus is after a killing frost and before the emergence of new shoots in the spring. Crop moisture in Illinois trials ranged from 50% in October to less than 10% by February.

The mineral content of harvested feedstock decreases with delayed harvest in both European and U.S. experience as rain and snow leach minerals from standing biomass. This reduction in mineral concentrations generally increases feedstock quality for conversion to biofuel. However, it comes at a price; harvestable biomass can decrease 30% to 50% over the winter as stems break with wind and weather.

Potential Yields (Tonnage and Energy Content)

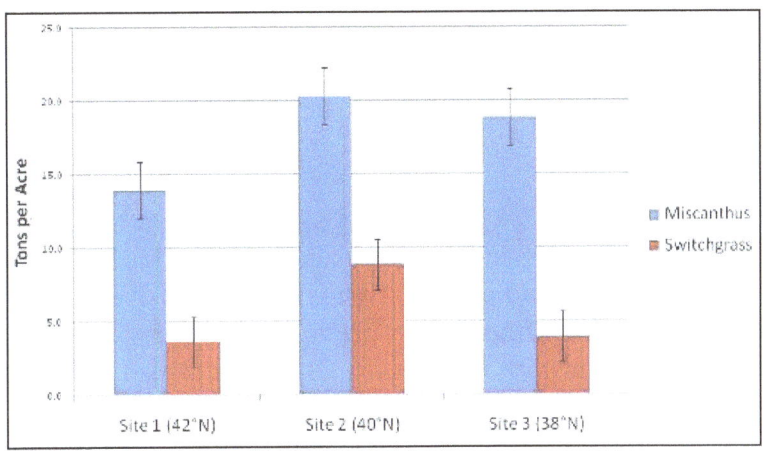

The average annual biomass yields of giant miscanthus and switchgrass harvested from three.

There is currently little published information on giant miscanthus yields in the United States. Tiller production and dry matter partitioning were examined in trials in Arkansas and varied little between giant miscanthus and energycane accessions. Similar measurements in Illinois showed giant miscanthus produced significantly fewer but larger tillers than switchgrass (Panicum virgatum) with increased tiller height and diameter contributing to greater biomass yields over switchgrass. Small plot trials of giant miscanthus have correlated well with modeled yield projections in Illinois and suggest seasonal peak biomass production of 12 to 20 tons per acre.

Additionally, numerous universities now have trials evaluating giant miscanthus, and thus the availability of yield data is expected to increase dramatically in the next few years.

In the absence of widespread observations in the United States, the European literature provides a good estimate of what may be expected from giant miscanthus under similar conditions. The considerable biomass data available from Europe suggest that giant miscanthus will yield from 10 to 40 Mg ha-1, with higher yields in warmer, wetter areas with moderately heavy soils. Further, the literature suggests that giant miscanthus is productive at high latitudes, e.g., 52 °N, or the equivalent of Hudson Bay. It remains to be seen whether giant miscanthus can withstand the cold temperatures prevalent in North America at these latitudes compared to the European equivalents.

Production Challenges

Propagation

Giant miscanthus has been propagated for research trials by digging plants from a field, then propagating them in a greenhouse. Giant miscanthus cannot be imported from Europe in any meaningful quantities due to current quarantine restrictions imposed by the USDA. Because giant miscanthus is a relative of sugarcane, it could conceivably harbor diseases that would threaten the U.S. sugarcane industry. Imported rhizomes must be monitored in quarantine greenhouses for three years before release, a costly process that effectively eliminates importation.

Overwinter Survival

A major bottleneck to giant miscanthus production in the upper Midwest is overwinter survival. Anecdotal evidence suggests that if a plant lives through the first winter, it is highly likely to survive subsequent winters. Cover during the winter also seems to affect survival: first-year stands that were mowed in October experienced high mortality in Michigan and Wisconsin compared to stands that were not mowed.

Estimated Production Cost

The costs of giant miscanthus production are front loaded. Depending on the source, planting material alone can cost $1,000 to $10,000 per acre. When considered over the productive lifetime of a stand, which is likely 15 to 20 years, the costs of production are less than annual row crops, leading to increased profitability, even without subsidy. Once established, giant miscanthus requires little maintenance, no annual replanting, and only an annual harvest. Given the low-input nature of the crop, it is likely that custom operators will arise to handle giant miscanthus production for interested land managers.

Environmental/Sustainability Issues

Because giant miscanthus has three sets of chromosomes and an uneven chromosome number, the chromosomes do not divide evenly during meiosis, leading to non-viable gametes, and hence to sterile seed. This is advantageous because it limits the capacity of giant miscanthus to spread unintentionally from seed, but it significantly complicates planting of new fields. In addition to lacking reproduction from seed, the rhizome structure of giant miscanthus spreads very slowly, thus minimizing vegetative spread. The oldest research stands in Europe were planted in the late 1980s and have only moved approximately 3 feet from their original location. Trials are currently under way in Illinois to evaluate risk of spread to and from agricultural lands.

Panicum Virgatum

Switchgrass (Panicum virgatum) is a native warm-season grass that is a leading biomass crop in the United States. More than 70 years of experience with switchgrass as a hay and forage crop suggest switchgrass will be productive and sustainable on rain-fed marginal land east of the 100th meridian. Long-term plot trials and farm-scale studies in the Great Plains and plot trials in the Great Plains, Midwest, South, and Southeast indicate switchgrass is productive, protective of the environment, and profitable for the farmer. Weed control is essential during establishment but with good management is typically not required again. Although stands can be maintained indefinitely, stands are expected to last at least 10 years, after which time the stand will be renovated, and new, higher-yielding material will be seeded on the site. Fertility requirements are well understood in most regions, with about 12 to 14 pounds of N per acre required for each ton of expected yield if the crop is allowed to completely senesce before the annual harvest.

Grassland scientists have conducted research on switchgrass (Panicum virgatum) for more than 70 years, with initial research focusing on livestock and conservation.

Recently, significant attention has been given to switchgrass as a model perennial grass for bioenergy production to reduce our dependence on foreign oil, boost our rural economies, reduce fossil fuel emissions, reduce erosion on marginal cropland, and enhance wildlife habitat.

Current Potential for use as a Biofuel

Switchgrass has excellent potential as a bioenergy feedstock for cellulosic ethanol production, direct combustion for heat and electrical generation, gasification, and pyrolysis. Switchgrass has several characteristics that make it a desirable biomass energy crop: it is a broadly adapted native to North America, it has consistently high yield relative to other species in varied environments, it requires minimal agricultural inputs, it is relatively easy to establish from seed, and a seed industry already exists.

Biology and Adaptation

Switchgrass.

Switchgrass is a perennial warm-season grass that is native to most of North America except for areas west of the Rocky Mountains and north of 55 °N latitude. Switchgrass grows 3 to 10 feet tall, typically as a bunchgrass, but the short rhizomes can form a sod over time. Switchgrass has high yield potential on marginal cropland and will be productive in most rain-fed production systems east of the 100th meridian. Productive switchgrass stands can be grown west of the 100th meridian with irrigation. Switchgrass is adapted to a wide range of habitats and climates and has few major insect or disease pests. Root depth of established switchgrass may reach 10 feet, but most of the root mass is in the top 12 inches of the soil profile. In addition to potential bioenergy production, switchgrass uses include pasture and hay production, soil and water conservation, carbon sequestration, and wildlife habitat.

Lowland vs Upland Ecotypes

Switchgrass has distinct lowland and upland ecotypes. Upland ecotypes occur in upland areas that are not subject to flooding, whereas lowland ecotypes are found on floodplains and other areas that receive run-on water. Generally, lowland plants have a later heading date and are taller with larger and thicker stems. Upland ecotypes are either octaploids or tetraploids, whereas lowland ecotypes are tetraploids. Lowland and upland tetraploids have been crossed to produce true F1 hybrids that have a 30 to 50% yield increase over the parental lines. These hybrids are promising sources for high-yielding bioenergy cultivars.

Production and Agronomic Information

Soybean stubble provides an excellent
seedbed for no-till seeding switchgrass.

During the establishment year, all harvests must occur after a killing frost to avoid damaging stands. In the establishment year, good weed management and rainfall will provide about half of the fully established yield potential of the site and cultivar.

Establishing Stands

Successful stand establishment during the seeding year is mandatory for economically viable switchgrass bioenergy production systems. Weed competition is the major reason for switchgrass stand failure. Acceptable switchgrass production can be delayed by one or more years by weed competition and poor stand establishment. Vogel and Masters reported a stand frequency of 50% or greater (two or more switchgrass plants per square foot) indicated a successful stand, whereas stand frequency from 25 to 50% was marginal to adequate, and stands with less than 25% frequency indicated a partial stand. In a study conducted on 12 farms in Nebraska, South Dakota, and North Dakota, switchgrass fields with a stand frequency of 40% or greater provided a successful stand.

Switchgrass is readily established when quality seed of an adapted cultivar is used with the proper planting date, seeding rate, seeding method, and weed control. In the central Great Plains, switchgrass can be planted two or three weeks before to two or three weeks after the recommended planting dates for corn (Zea mays), typically from late April to early June. Switchgrass should be seeded at 30 pure live seed (PLS) per square foot (5 PLS pounds per acre) based on the quality of the seedlot. Excellent results are obtained by planting after a soybean (Glycine max) crop using a properly calibrated no-till drill with depth bands that plant seeds 0.25 inch to 0.5 inch deep followed by press wheels. Row spacing for switchgrass is typically 7.5 to 10 inches. If switchgrass is planted after crops that leave heavy residue such as corn or sorghum (Sorghum bicolor), it may be necessary to graze the residue, shred or bale the stalks, or use tillage to reduce the residue. If tillage is required, the seedbed needs to be packed to firm the soil. The packed soil needs to be firm enough so that walking across the field leaves only a faint footprint. Applying 8 oz of quinclorac plus 1 qt of atrazine per acre immediately after planting has provided effective grassy and broadleaf weed control for establishment. The most cost-effective method to control broadleaf weeds in switchgrass fields during the establishment year is to apply 2,4-D at 1 to 2 qt acre-1 after switchgrass seedlings have about four leaves. After the establishment year, a successfully established switchgrass stand requires limited herbicide applications.

Seeding into corn or sorghum stubble may require
plowing, disking, and packing to develop a firm seedbed.

Pack the tilled soil until walking across the field leaves only a faint footprint to ensure good seed-to-soil contact and prevent soil in-filling of the packer wheel depression.

Nitrogen (N) fertilizer is not recommended during the planting year since N will encourage weed growth, increase competition for establishing seedlings, increase establishment cost, and increase economic risk associated with establishment if stands should fail. Soil tests are recommended prior to planting. Since switchgrass is deep rooted, soil samples should be taken from each 1-foot increment to a depth of 5 feet. In most agricultural fields, adequate levels of phosphorus (P) and potassium (K) will be in the soil profile. If warranted by soil tests, P and K can be applied before seeding to encourage root growth and promote rapid establishment. Recommended P levels for the western corn belt are in table. Switchgrass can tolerate moderately acidic soils, but optimum seed germination occurs when soil pH is between 6 and 8. With good weed management and favorable precipitation, a crop equal to about half of potential production can be harvested after frost at the end of the planting year, with 75 to 100% of full production achieved the year after planting.

Table: Phosphorus (P) recommendations for the western corn belt based on two common soil test levels.

P Index Value	Soil Test Levels		P Rate
	Bray & Kurtz #1	Olsen P (Na HCO$_3$)	
	——————— ppm ———————		lb P$_2$O$_5$/Acre
Very low	0-5	0-3	40
Low	6-15	4-7	20
Medium	16-25	8-14	10
High	25+	15+	0

Established Stands

Although switchgrass can survive on low fertility soils, it does respond to fertilizer, especially N. The amount of N required by switchgrass is a function of the yield potential of the site, productivity of the cultivar, and other management practices being used. Consequently, the optimum N rate for switchgrass managed for biomass will vary, but a few references indicative of the responses to N in different regions of the United States are included. Additionally, biomass will decline over years if inadequate N is applied, and yield will be sustainable only with proper N application. In Nebraska and Iowa, Cave-in-Rock yield increased as N rate increased from 0 to 270 lb N acre-1, but soil N increased when more than 100 lb N acre-1 were applied. They reported biomass was optimized by applying 100 lb N acre-1, with about the same amount of N being applied as was being removed by the crop. A general N fertilizer recommendation for the Great Plains and Midwest region is to apply 20 lb N acre-1 yr-1 for each ton of anticipated biomass if harvesting during the growing season, with N rate reduced to 12 to 14 lb N acre-1 yr-1 for each ton of anticipated biomass if harvesting after a killing frost. The N rate can be reduced when the harvest is after a killing frost because switchgrass cycles some N back to roots during autumn. If soil tests indicate a new switchgrass field has high residual N levels, N rates can be significantly reduced during the initial production years using the above information as a guideline. Apply N at switchgrass green-up to minimize cool-season weed competition.

Table: Switchgrass publications addressing nitrogen fertilizer application for different regions of the United States

State(s)	Parameters Evaluated
AL	N rate and row spacing effect on C partitioning
IA	Yield and quality parameters for 20 strains
IA, NE	Harvest date and N rate effects
NC, KY, TN, VA, WV	Long-term yield under different management regimes
SD	Harvest date and N rate effects on biomass, persistence, species composition, and soil organic carbon of switchgrass-dominated CRP
TX	Yield and stand responses to N and P as affected by row spacing

Spraying herbicides to control broadleaf weeds typically is needed only once or twice every 10 years in an established, well-managed switchgrass stands. When needed, the most effective and economical approach is with broadcast applications of 2,4-D at 1 to 2 qt acre-1. Spray broadleaf weeds as early in the growing season as possible to reduce the impact of weed interference on switchgrass yield. In some cases, cool-season grasses may invade switchgrass stands and reduce yield. Harvesting after switchgrass senescence in autumn but while cool-season grasses are growing, then applying glyphosate at 1 to 2 qt acre-1, is an effective method to reduce cool-season grasses. However, make certain switchgrass is dormant when glyphosate is applied, or stands could be damaged. Spring applications of atrazine at 2 qt acre-1 can be used to control cool-season grasses in established switchgrass stands.

Rotary head mowers (disc mowers) effectively harvested
this 6-ton per acre switchgrass field at anthesis.

Additionally, after a killing frost, the multidirectional arrangement of the switchgrass in the windrow was easier to bale than the linearly arranged windrow left by the sickle-bar head.

Harvest and Storage

Maximizing yield currently is the primary objective when harvesting biomass feedstocks. In the Great Plains and Midwest, maximum first-cut yields are attained by harvesting switchgrass when panicles are fully emerged to the post-anthesis stage. Sufficient regrowth may occur about one year out of four to warrant a second harvest after a killing frost. Do not harvest switchgrass within six weeks of the first killing frost or shorter than a 4-inch stubble height to ensure translocation of storage carbohydrates to maintain stand productivity and persistence. Dormant season harvests after a killing frost will not damage switchgrass stands but will reduce the amount of snow captured during winter. In general, a single harvest during the growing season maximizes switchgrass biomass recovery, but harvesting after a killing frost will ensure stand productivity and persistence, especially when drought conditions occur, and reduce N fertilizer requirements. Delaying harvest until spring will reduce moisture and ash contents, but yield loss can be as high as 40% compared with a fall harvest. With proper management, productive stands can be maintained indefinitely and certainly for more than 10 years. Harvesting switchgrass in summer at or after flowering when drought conditions exist is not recommended.

Switchgrass can be harvested and baled with commercially available haying equipment. Self-propelled harvesters equipped with a rotary head (disc mowers) have most effectively harvested high-yielding (>6-ton per acre) switchgrass fields. Additionally, after a killing frost, the multidirectional arrangement of the switchgrass in the windrow was easier to bale than the linearly arranged windrow left by a sickle-bar head. Round bales tend to have less storage losses than large square bales (>800 lb) when stored outside, but square bales tend to be easier to handle and load a truck for transport without road width restrictions. After harvest, poor switchgrass storage conditions can result in storage losses of 25% in a single year. In addition to storage losses in weight, there can be significant reductions in biomass quality, and the biomass may not be in acceptable condition for a biorefinery. Switchgrass grown for use in a biorefinery may have to be stored for a full year or longer since biorefineries will operate 365 days a year. Some type of covered storage will be necessary to protect the producer's investment.

Proper storage of switchgrass bales is imperative to maintain total harvested dry matter and prevent spoilage.

Large square bales can spoil from the top and bottom and can lose more than 25% of total dry matter in six months when stored outside in the open (top left), but covering the large square bales with hay tarps (top right) reduces dry matter loss to about 7% in six months. Wrapping big round bales with at least three wraps of net-wrap maintains the structure of the bale and reduces the surface area of the bale that contacts the ground. Covering big round bales stored outside can reduce dry matter loss to less than 3% in six months.

Potential Yield

Switchgrass yield is strongly influenced by precipitation, fertility, soil, location, genetics, and other factors. Most plot and field-scale switchgrass research has been conducted on forage-type cultivars selected for other livestock-based characteristics in addition to yield. Consequently, the forage-type cultivars in the Great Plains and Midwest are entirely represented by upland ecotypes which are inherently lower yielding than lowland ecotypes. Thus, yield data comparing forage-type upland cultivars like Cave-In-Rock, Shawnee, Summer, and Trailblazer do not capture the full yield potential of switchgrass and are not fair comparisons. For example in Nebraska, high-yielding F1 hybrids of Kanlow and Summer produced 9.4 tons acre-1 year-1, which was 68% greater than Summer and 50% greater than Shawnee. New biomass-type switchgrass cultivars will be available in the near future for the Great Plains and Midwest. Knowing the origin of a switchgrass cultivar is important since switchgrass is photoperiod sensitive. Planting a switchgrass cultivar too far north of the cultivar origin area (>300 miles) can result in winter stand loss. Planting a switchgrass cultivar south of its origin area results in less biomass because the shorter photoperiod causes plants to flower too early.

Production Challenges

There are major challenges to using switchgrass for cellulosic ethanol. An ethanol plant requires a reliable and consistent feedstock supply. A 50-million-gallon per year plant will require 625,000 U.S. tons of feedstock per year assuming 80 gallons of ethanol can be produced from one ton of feedstock. Although cellulosic ethanol plants likely will use multiple feedstocks, this example assumes switchgrass will be the only feedstock. Operating every day of the year, the plant will require 1,712 dry matter (DM) tons of feedstock per day, or 342 acres of switchgrass yielding 5 DM tons per acre. If a loaded semi can deliver 30 round bales each containing 0.6 DM tons (18 U.S. tons), the ethanol plant will use 95 semi loads of feedstock per day, requiring a semi to be unloaded every 15 minutes 24 hours per day, 7 days per week.

There must be an available land base in the local agricultural landscape to produce feedstock. The biomass and ethanol yield of the feedstock will determine the land area required for feedstock production. Assuming 25 miles is the maximum economically feasible distance feedstock can be transported, all of the feedstock must be grown within a 25-mile radius of the bio-refinery, an area containing about 1.26 million acres. Assuming a 50-million-gallon per year cellulosic ethanol plant requires 625,000 tons of feedstock per year, if feedstock yield is 1.75 DM tons/acre, 28% of the land would need to grow the feedstock, and this is not feasible in most agricultural areas. At 5 DM tons/acre, a commonly-achieved yield with available forage cultivars, only 10% of the land would be needed for feedstock production and is feasible in most agricultural areas. However, at 10 DM tons/acre, only 5% of the land base would be needed for feedstock production and would minimally alter the agricultural landscape. Dry matter yield will exceed 10 tons/acre in many areas of the South and Southeast, so less than 5% of the land base would be needed for feedstock production. This example reinforces the importance of high DM yield to the agricultural feasibility of cellulosic ethanol, not to mention the inability of the producer to profit by growing low-yielding energy crops. A majority of the switchgrass likely will be grown on marginal lands that have suboptimal characteristics (i.e., slope, soil depth, etc.) for producing food and feed, or on lands currently enrolled in conservation programs.

Table. Reported dry matter (DM) yield, acres required to grow 625,000 tons of dry matter per year, and the percent of the land base required to provide feedstock for a 50-million-gallon cellulosic ethanol plant for different herbaceous perennial feedstocks in the Great Plains and Midwest.

Feedstock	Yield, DM tons/acre	Acres needed to grow 625,000 DM tons/year	Percent of land in 25-mile radius
LIHD prairie	1.75	357,000	28
Managed native prairie	2.5	250,000	20
Shawnee switchgrass	5	125,000	10
Bioenergy switchgrass	7.4	84,460	6.6
Hybrid switchgrass	9.4	66,489	5.3

Growing switchgrass must be profitable for the producer, it must fit into existing farming operations, it must be easy to store and deliver to the ethanol plant, and extension efforts must be provided to inform producers on the agronomics and best management practices for specific regions, all of which have been addressed for switchgrass. Switchgrass fits well into the production systems of most farmers. Harvesting switchgrass after frost is a time when most farmers have completed corn and soybean harvests and handling switchgrass as a hay crop is not foreign to most producers. The economic opportunities of switchgrass for small, difficult-to-farm, or poorly-productive fields will be attractive to many producers.

There are potential difficulties with large-scale switchgrass monocultures, but most are speculation at this point. Concerns arise for potential disease and insect pests, and the escape of switchgrass as an invasive species with the production of millions of switchgrass acres, especially since little research has been conducted on these topics. Most pathogen issues cannot be fully realized until large areas are planted to switchgrass. However, the broad genetic diversity available to switchgrass breeders, the initial pathogen screening conducted during cultivar development, and the fact that switchgrass has been a native component of central U.S. grasslands for centuries will likely limit the negative pest issues. Switchgrass has been used widely throughout the Great Plains and Midwest for pasture and conservation purposes for decades, and no invasive problems have developed or been identified.

Environmental and Sustainability Issues

Sustainable biomass energy crops must be productive, protective of soil and water resources, and profitable for the producer. Numerous studies have reported that switchgrass will protect soil, water, and air quality; provide fully sustainable production systems; sequester C; create wildlife habitat; increase landscape and biological diversity; return marginal farmland to production; and increase farm revenues. Switchgrass root density in the surface 6 inches is two-fold greater than alfalfa, more than three-fold greater than corn, and more than an order of magnitude greater than soybean. In a five-year field study conducted on 10 farms in Nebraska, South Dakota, and North Dakota, Liebig et al. reported that switchgrass stored large quantities of C, with four farms in Nebraska storing an average of 2,590 pounds of soil organic C (SOC) acre-1 year-1 when measured to a depth of 4 feet. However, they noted that SOC increases varied across sites, and the variation in SOC change reiterated the importance of long-term environmental monitoring sites in major agro-ecoregions.

Energy produced from renewable carbon sources is held to a different standard than energy produced from fossil fuels, in that renewable fuels must have highly positive energy values and low greenhouse gas emissions. The energy efficiency and sustainability of ethanol produced from grains and cellulosics has been evaluated using net energy value (NEV), net energy yield (NEY), and the ratio of the biofuel output to petroleum input [petroleum energy ratio (PER)]. An energy model using estimated agricultural inputs and simulated yields predicted switchgrass could produce greater than 700% more output than input energy. These modeled results were validated with actual inputs from multi-farm, field-scale research to predict energy output. Switchgrass fields on 10 farms in Nebraska, South Dakota, and North Dakota produced 540% more renewable energy (NEV) than nonrenewable energy consumed over a five-year period. The estimated on-farm NEY was 93% greater than human-made prairies and 652% greater than low-input switchgrass grown in small plots in Minnesota. The 10 farms and five production years had a PER of 13.1 and produced 93% more ethanol per acre than human-made prairies and 471% more ethanol per acre than low-input switchgrass in Minnesota. Average greenhouse gas (GHG) emissions from switchgrass-based ethanol were 94% lower than estimated GHG emissions for gasoline. Switchgrass for bioenergy is an energetically positive and environmentally sustainable production system for the Great Plains.

Implementing switchgrass-based bioenergy production systems will require converting marginal land from annual row crops to switchgrass and could exceed 10% in some regions depending on the yield potential of the switchgrass strains. In a five-year study in Nebraska, the potential ethanol yield of switchgrass averaged 372 gallons acre-1 and was equal to or greater than that for no-till corn (grain + stover) on a dry-land site with marginal soils. Removing 50% of the corn stover each year reduced subsequent corn grain yield, stover yield, and total biomass. Growing switchgrass on marginal sites likely will enhance ecosystem services more rapidly and significantly than on more productive sites.

Feasibility

Perennial herbaceous energy crops provide several challenges. A stable and consistent feedstock supply must be available year-round to the ethanol or power plant. For the producer, perennial herbaceous energy crops must be profitable, they must fit into existing farming operations, they must be easy to store and deliver to the plant, and extension efforts must be provided to inform producers on the agronomics and best management practices for growing perennial herbaceous energy crops. However, perennial herbaceous energy crops have potential for improvement, and they present a unique opportunity for cultural change on the agricultural landscape. There are numerous environmental benefits to perennial herbaceous cropping systems that can improve agricultural land use practices such as stabilizing soils and reducing soil erosion, improved water quality, increased and improved wildlife habitat, and storing C to mitigate greenhouse gas emissions. There is large potential for achieving all of these benefits, provided agronomic, genomic, and operational aspects of perennial herbaceous cropping systems are fully developed and accepted by farmers. Herbaceous perennial energy crops may be used in conjunction with agriculture residues (corn stover and wheat straw), which likely would be harvested in autumn, and perennial grasses could be harvested in very early spring while they are dry, similar to when prairies are typically burned. This may help reduce the need for feedstock storage by providing feedstock at different times during the year.

Growing seed to meet potential demand for bioenergy will not be an issue. Switchgrass has many desirable seed characteristics and can produce viable seed during the seeding year, especially under irrigation. Established seed production fields can produce 500 to 1,000 pounds of seed per acre with irrigation, and the seed is easily threshed, cleaned, and planted with commercial planting equipment. Seed production systems are well established, and a commercial industry for switchgrass seed has existed for over 50 years.

ALGAE-BASED FEEDSTOCKS

In recent years, biofuel production from algae has attracted the most attention among other possible products. This can be explained by the global concerns over depleting fossil fuel reserves and climate change. Furthermore, increasing energy access and energy security are seen as key actions for reducing poverty thus contributing to the Millennium Development Goals. Access to modern energy services such as electricity or liquid fuels is a basic requirement to improve living standards.

One of the steps taken to increase access and reduce fossil fuel dependency is the production of bio-fuels, especially because they are currently the only short-term alternative to fossil fuels for transportation, and so until the advent of electromobility. The so-called first generation biofuels are produced from agricultural feedstocks that can also be used as food or feed purposes. The possible competition between food and fuel makes it impossible to produce enough first generation biofuel to offset a large percentage of the total fuel consumption for transportation. As opposed to land-based biofuels produced from agricultural feedstocks, cultivation of algae for biofuel does not necessarily use agricultural land and requires only negligible amounts of freshwater (if any), and therefore competes less with agriculture than first generation biofuels. Combined with the promise of high productivity, direct combustion gas utilization, potential wastewater treatment, year-round production, biochemical content of algae and chemical conditions of their oil content can be influenced by changing cultivation conditions. Since they do not need herbicides and pesticides, algae appear to be a high potential feedstock for biofuel production that could potentially avoid the aforementioned problems. On the other hand, microalgae, as opposed to most plants, lack heavy supporting structures and anchorage organs which pose some technical limitations to their harvesting. The real advantage of microalgae over plants lies in their metabolic flexibility, which offers the possibility of modification of their biochemical pathways (e.g. towards protein, carbohydrate or oil synthesis) and cellular composition. Algae-based biofuels have an enormous market potential, can displace imports of fossil fuels from other countries (hence reduce a country's dependence), and is one of the new, sustainable technologies which can count on ever-increasing political and consumer support.

The reasons for investigating algae as a biofuel feedstock are strong but these reasons also apply to other products that can be produced from algae. There are many products in the agricultural, chemical or food industry that could be produced using more sustainable inputs and which can be produced locally with a lower impact on natural resources. Co-producing some of these products together with biofuels, can make the process economically viable, less dependent from imports and fossil fuels, locally self sufficient and expected to generate new jobs, with a positive effect on the overall sustainability.

A wave of renewed interest in algae cultivation has developed over the last few years. Scientific research, commercialization initiatives and media coverage have exploded since 2007. In most cases, the main driver of the interest in algae is its high potential as a renewable energy source, mainly algae-based biofuels (ABB) for the transport sector.

Cultivation Systems for Algae

Although not specific to biofuel production from algae, it is important to understand the basics of algae cultivation systems. Systems which use artificial light demand, per definition, more energy in lighting than what is gained as algal energy feedstock.

Seaweed has historically been harvested from natural populations or collected after washing up on shore. To a much lesser extent, a few microalgae have also been harvested from natural lakes by indigenous populations. Given that these practices are unlikely to sustain strong growth, only the cultivation of algae in man-made systems will be considered.

A production system is geared towards a high yield per hectare because it reduces the relative costs for land, construction materials and some operation costs. It is not uncommon for published yield estimates to be too high, sometimes higher than theoretically possible. These overestimations lead to unrealistic expectations. Realistic estimates for productivity are in the order of magnitude of 40-80 tons of dry matter per year per hectare, depending on the technology used and the location of production. This is still substantially higher than almost all agricultural crops. Surpassing yields of 80 tons per year per hectare will likely require genetically improved strains or other technologies able to counteract photosaturation and photoinibition.

Open Cultivation Systems

The main large-scale algae cultivation system is the so-called raceway pond. These are simple closed-loop channels in which the water is kept in motion by a paddle wheel. The channel is usually 20-30 cm deep and made of concrete or compacted earth, often lined with white plastic. It is designed for optimal light capture and low construction costs. The main land requirement is that of flat land.

Process control in such an open system is difficult since these are unstable ecosystems, temperature is dependent on the weather and, depending on climatic conditions, large amounts of water cyclically evaporate or are added by rainfall. Furthermore, the open character of the system makes it possible for naturally occurring algae or algae predators to infiltrate the system and compete with the algae species intended to be cultivated. Therefore a monoculture can only be maintained under extreme conditions, like high salinity (e.g. Dunaliella), high pH (e.g. Spirulina) or high nitrogen (e.g. Chlorella) water. These conditions generally limit optimal growth and operate at a low algae concentration, making harvesting more difficult.

In conclusion, there is an important trade-off between a low price for the cultivation system and its production potential.

Closed Cultivation Systems

Many of the problems of open systems can be mitigated by building a closed system which is less influenced by the environment. Many configurations exist but all of them rely on the use of transparent plastic containers (usually tubes) through which the culture medium flows and in which the algae are exposed to light. Such a system is clearly more expensive and therefore capital intensive if produced on a large scale, but allows a wider number of species to be cultivated under ad-hoc conditions, normally with a higher concentration and productivity. On the other hand, these systems suffer from high energy expenditures for mixing and cooling than open ponds and are also technically more difficult to build and maintain.

Closed systems allow for the cultivation of algal species that cannot be grown in open ponds.

Sea-based Cultivation Systems

Whereas the previously described cultivation systems are almost exclusively used for microalgae, algae cultivation in the sea is the domain of seaweed. Seaweed cultivation, although very labour intensive near shore in shallow water and often at small-scale, is common practice in parts of Asia. To make an impact as bioenergy feedstock, seaweed should be produced in floating cultivation systems spanning hundreds of hectares. Most seaweeds require a substrate to hook to; which in practice means that the cultivation system must contain a network of ropes. The amount of construction material could be drastically reduced when free-floating seaweed (like some Sargassum species) is cultivated as just a structure to contain the colony would then be needed. Sea-based systems are less well developed than land based systems, although some R&D initiatives have been undertaken and are still ongoing. The system for seaweed cultivation in China has not changed much since it was invented in the 1950s, although options for modernization have been identified. Some countries, such as Chile, are important seaweed producers, but rely completely on the harvesting of natural populations.

Algae-based Bioenergy Products

There are a variety of ways to produce biofuel with algae. Figure provides an overview of the options.

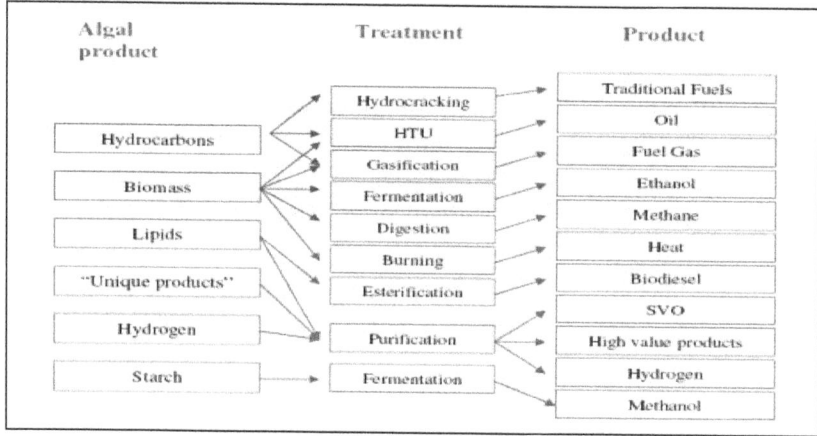

Overview of a Igae-to- energy options.

Biodiesel

Biodiesel production from algal oils has received most attention since algae can contain potentially over 80% total lipids, (while rapeseed plants, for instance, contain about 6% lipids). Under normal growth conditions the lipid concentration is lower (<40%) and high oil content is always associated with very low yields. The various lipids production can be stimulated under stress conditions, e.g. insufficient nitrogen availability. Under such conditions, biomass production is not optimal though, reducing the non-lipid part of the biomass that can be further used as a source for co-products.

Hydrocarbons

One genus of algae, Botryococcus, does not produce the above-mentioned lipids, but longer chain hydrocarbons, which are not suitable for biodiesel production. Instead, they can be converted in a process similar to the production of conventional fuels from fossil oil. Botryococcus is a freshwater species but can also grow in saline water and it can produce certain carotenoids. Its drawback is the relatively slow growth speed.

Ethanol

Ethanol is commonly produced from starch-containing feedstocks; some algae have been reported to contain over 50% of starch. Algal cell walls consist of polysaccharides which can be used as a feedstock in a process similar to cellulosic ethanol production, with the added advantage that algae rarely contain lignin and their polysaccharides, are generally more easily broken down than woody biomass. Co-products can potentially be derived from the non-carbohydrate part of the algal biomass.

Biogas

Anaerobic digestion converts organic material into biogas that contains about 60%-70% biomethane, while the rest is mainly CO_2, which can be fed back to the algae. A main advantage is that this process can use wet biomass, reducing the need for drying. Another advantage is that the nutrients contained in the digested biomass can be recovered from the liquid and solid phase.

Biogas as the main product is not economically viable, but this process can be applied to any left-over biomass after extraction of a co-product.

Thermochemical Treatment

The biological treatment of algal organic material has a non-biological counterpart, with the advantage that no live organisms are involved and therefore more varied and extreme process conditions can be used. The biomass undergoes a chemical conversion under high temperature and pressure conditions. Depending on the water content and how extreme these conditions are, the biomass carbon ends up in a raw gaseous, liquid or solid phase which can be upgraded for usage as a biofuel.

The energy input of this type of treatment is clearly higher compared to biogas production.

Hydrogen

Some algae can be manipulated into producing hydrogen gas. Currently the yield of this process is low and since energy is lost by the cells to form hydrogen, not much biomass is produced and therefore there is little potential for co-production.

Bioelectricity

Algal biomass can also be co-combusted in a power plant. For this, the biomass needs to be dried, which implies a significant amount of energy. This process is thus only interesting if the biomass is required to be dried in order to extract a certain co-product as a first step before being used as a biofuel.

References

- Corn-for-biofuel-production: farm-energy.extension.org, Retrieved 20 July, 2019

- Gad loebenstein; george thottappilly (2009). The sweetpotato. Pp. 391–425. Isbn 978-1-4020-9475-0

- Cassava-residues-could-provide-sustainable-bioenergy-for-cassava-producing-nations, cassava: intechopen. com, Retrieved 31 July, 2019

- "nahuatl influences in tagalog". El galéon de acapulco news. Embajada de méxico, filipinas. Archived from the original on 27 april 2013. Retrieved 16 february 2012

- Soybeans-for-biodiesel-production: farm-energy.extension.org, Retrieved 17 February, 2019

- Rapeseed-and-canola-for-biodiesel-production: farm-energy.extension.org, Retrieved 21 March, 2019

- Palm-oil-biodiesel-a-preferred-biofuel-feedstock: palmoiltoday.net, Retrieved 16 June, 2019

- Jatropha-biofuel-industry-the-challenges, frontiers-in-bioenergy-and-biofuels: intechopen.com, Retrieved 28 February, 2019

- Switchgrass-panicum-virgatum-for-biofuel-production: farm-energy.extension.org, Retrieved 18 April, 2019

Biofuels

The fuel which is produced using modern techniques and processes from biomass is referred to as biofuel. Some of the various types of biofuels are algal fuel, methanol fuel, ethanol fuel, butanol fuel, pellet fuel, biodiesel, etc. All these diverse types of biofuels have been carefully analyzed in this chapter.

GENERATIONS OF BIOFUEL

Biofuels are energy sources made from recently grown biomass (plant or animal matter). Biofuels have been around for a long time, but petroleum and coal have been used primarily as energy sources due to their high abundance, high energy value, and cheap prices. Fossil fuels such as coal and petroleum also come from biomass but the difference is that they took millions of years to produce. Biofuels are making a resurgence due to increasing oil prices, dwindling fossil fuel reserves, the desire to have a renewable, reliable source of energy and as a way to mitigate the effects of climate change. Biofuels are a renewable resource because they are continually replenished. Fossil fuels on the other hand are not renewable since they require millions of years to form.

There are three types of biofuels: 1st, 2nd and 3rd generation biofuels. They are characterized by their sources of biomass, their limitations as a renewable source of energy, and their technological progress. The main drawback of 1st generation biofuels is that they come from biomass that is also a food source. This presents a problem when there is not enough food to feed everyone. 2nd generation biofuels come from non-food biomass, but still compete with food production for land use. Finally, 3rd generation biofuels present the best possibility for alternative fuel because they don't compete with food. However, there are still some challenges in making them economically feasible.

First generation biofuels, also known as conventional biofuels, are made from sugar, starch or vegetable oil. First generation biofuels are produced through well-understood technologies and processes, like fermentation, distillation and transesterification. These processes have been used for hundreds of years in many uses, such as making alcohol. Sugars and starches are fermented to produce primarily ethanol, and in smaller quantities, butanol and propanol. Ethanol has one-third of the energy density of gasoline, but is currently used in many countries, including the United States, as an additive to gasoline. A benefit of ethanol is that it burns cleaner than gasoline and therefore produces less greenhouse gases. Another 1st generation biofuel, called biodiesel, is produced when plant oil or animal fat goes through a process called transesterification. This process involves exposing oils with an alcohol such as methanol in the presence of a catalyst. The distillation process involves separating the main product from any of the by-products of the reactions.

Biodiesel can be used in place of petroleum diesel in many diesel engines or in a mixture of the two. The figure shows the process of how 1st generation biofuels are made.

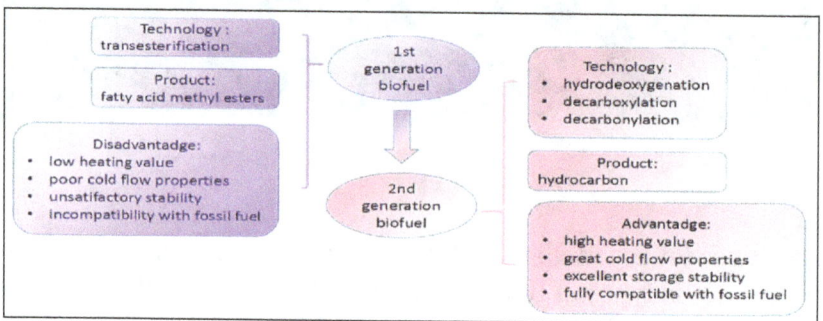

First Generation Biofuels

First generation biofuels symbolize a step towards energy independence and weaning off fossil fuels for energy demands. These biofuels also support agricultural industries and rural communities through increased demand for crops. This being said, 1st generation biofuels also have several disadvantages. They pose a threat to food prices since the biomass used are food crops such as corn and sugar beet. First generation biofuel production has contributed to recent increases in world prices for food and animal feeds. They also have the potential to have a negative impact on biodiversity and competition for water in some regions. Additionally, biomass for first generation biofuels requires lots of land to grow, and this provides only limited greenhouse gases reduction. They also only provide a small benefit over fossil fuels in regards to greenhouse gases since they still require high amounts of energy to grow, collect, and process. Current production practices use fossil fuels for power. First generation biofuels are also a more expensive option than gasoline, making it economically unfavorable. Finally, biodiesel almost always comes from recycled oils from restaurants, as opposed to virgin oils, so the supply is limited by restaurants' oil use.

Second Generation Biofuels

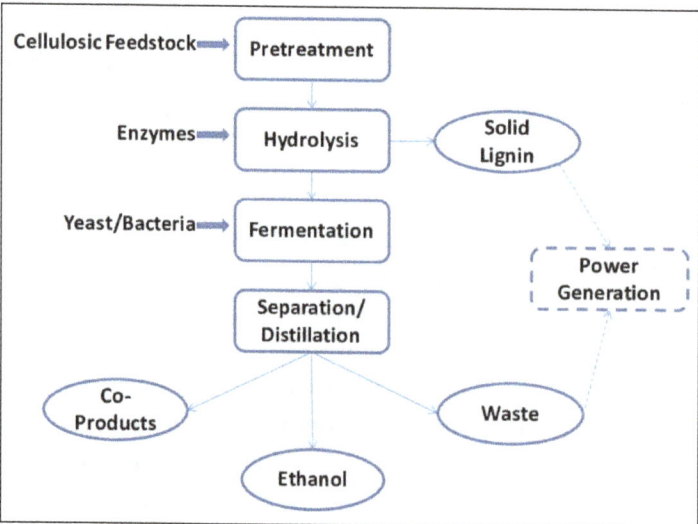

The biomass sources for 2nd generation biofuels include wood, organic waste, food waste and specific biomass crops. Fast growing trees such as poplar trees need to undergo a pretreatment step, which is series of chemical reactions that break down lignin, the "glue" that holds plants together, in order to make fuel. This pretreatment step involves thermochemical or biochemical reactions that unlock the sugars embedded in fibers of the plant. After this step, the process to generate plant ethanol resembles that of 1st generation ethanol production. Additionally, straw and other forest residues can go through a thermochemical step that produces syngas (a mixture of carbon monoxide, hydrogen and other hydrocarbons). Hydrogen can be used as a fuel and the other hydrocarbons can be used as additives to gas oil. The Figure shows how 2nd generation biofuels are produced.

Second generation biofuels address many issues associated with 1st generation biofuels. They don't compete between fuels and food crops since they come from distinct biomass. Second generation biofuels also generate higher energy yields per acre than 1st generation fuels. They allow for use of poorer quality land where food crops may not be able to grow. The technology is fairly immature, so it still has potential of cost reductions and increased production efficiency as scientific advances occur. However, some biomasses for second-generation biofuels still compete with land use since some of the biomass grows in the same climate as food crops. This leaves farmers and policy makers with the hard decision of which crop to grow. Cellulosic sources that grow alongside food crops could be used for biomass, such as corn stover (leaves, stalk, and stem of corn). However, this would take away too many nutrients from the soil and would need to be replenished through fertilizer. In addition, the process to produce 2nd generation fuels is more elaborate than 1st generation biofuels because it requires pretreating the biomass to release the trapped sugars. This requires more energy and materials.

Third Generation Biofuels

Third generation biofuels use specially engineered crops such as algae as the energy source. These algae are grown and harvested to extract oil within them. The oil can then be converted into biodiesel through a similar process as 1st generation biofuels, or it can be refined into other fuels as replacements to petroleum-based fuels. The figure shows the general steps of how 3rd generation biofuels are produced.

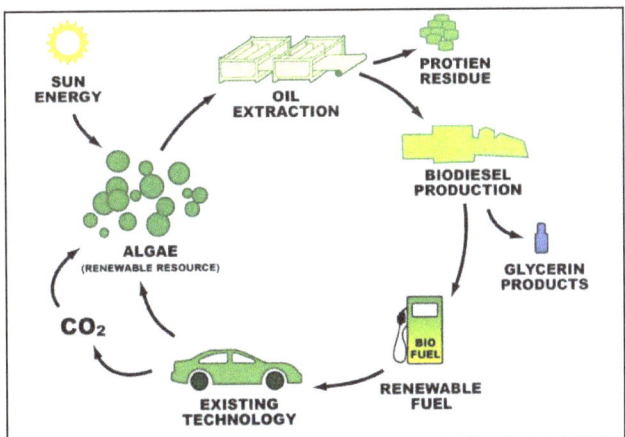

Third generation biofuels are more energy dense than 1st and 2nd generation biofuels per area of harvest. They are cultured as low-cost, high-energy, and completely renewable sources of energy. Algae are advantageous in that it can grow in areas unsuitable for 1st and 2nd generation crops, which would relieve stress on water and arable land used. It can be grown using sewage, wastewater, and saltwater, such as oceans or salt lakes. Because of this, there wouldn't be a need to use water that would otherwise be used for human consumption. However, further research still needs to be done to further the extraction process in order to make it financially competitive to petrodiesel and other petroleum-based fuels.

Market Barriers of Biofuels

The promotion of renewable energies is faced by various market barriers. These barriers limit the development of renewables unless special policy measures are enacted, unless no other fossil resources are available or unless the price advantage of renewables highly exceeds that of fossil fuels. In order to promote a fast introduction of biofuels, barriers have to be detected and solutions have to be found.

The Union of Concerned Scientists has formulated four main categories of barriers to the use of renewable energy technologies (RET) in general:

- Commercialization barriers faced by new technologies competing with mature technologies.

- Price distortions from existing subsidies and unequal tax burdens between renewables and other energy sources.

- Failure of the market to value the public benefits of renewables.

- Other market barriers such as inadequate information, lack of access to capital, high transaction costs.

These barriers to RETs also apply to biofuels. In order to find solutions for overcoming these barriers, they have to be described in more detail. The main market constraints specific to biofuels can be summarized by nine main market barriers:

- Economical barriers: The production of biofuels is still expensive, markets are immature and beneficial externalities are not accounted.

- Technical barriers: The fuel quality is not yet constant and conversion technologies for certain biofuels are still immature (e.g. for synthetic biofuels).

- Trade barriers: For some biofuels still no quality standards exist. Also no common European sustainability standard exists. Barriers exist for international trade of bioethanol due to denaturation obligations.

- Infrastructural barriers: Depending on the type of biofuel, new or modified infrastructures are needed. Especially the use of biohydrogen and biomethane need profound infrastructural changes.

- Causality dilemma: Before owners of filling stations sell biofuels, they claim that car manufacturers have to sell refitted cars first. The automotive industry claims that the infrastructure has to be developed first. This dilemma is a visible barrier for the introduction of FFV and the promotion of E85 in some European countries.

- Ethical barriers: Biomass feedstock sources may compete with food supply.

- Knowledge barriers: The general public, but also decision makers and politicians are lacking knowledge on biofuels.

- Political barriers: Lobbying groups influence politicians to create or conserve an unfavorable political framework for biofuels.

- Conflict of interest: Conflict between 'promoters' of first and second generation biofuels may weaken the overall development of biofuels.

Above mentioned barriers will also largely depend on the type of biofuel and the specific framework conditions. In the following years significant technological promotional and political challenges are thus to be faced in order to establish biofuel as a main pillar of a sustainable worldwide transportation system.

BIOFUEL LIFE CYCLE

Biofuels can have positive or negative impacts on various issues. In order to assess benefits from the utilization of biofuels compared to fossil fuels, life cycles have to be determined. Life cycles largely depend on type of feedstock, choice of location, production of by- products, process technology and on how the fuel is used. Within this variety, the basic components of life cycles in biofuel processing are always the same. Therefore some aspects of the general life cycle of biofuels are presented.

As it is shown in figure the life cycle of biofuels has several vertical process steps: biomass production and transport, biofuel processing, biofuel distribution and biofuel consumption. In addition, the industrial process steps of creating fertilizers, seeds and pesticides for the production of biomass must be included.

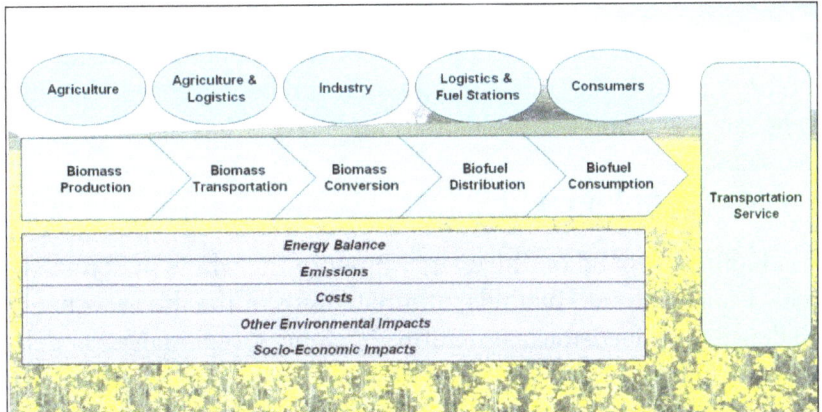

Actors, life cycle and horizontal attributes of biofuel production.

In each process step of biofuel production different actors are involved. Biomass is produced and transported by farmers. It is sometimes also transported by logistic services or by the biomass conversion industry itself. The conversion of biomass to biofuels can be either made by farmers or by industry, which is more common. Finally, biofuels are distributed by logistic services or fuel stations and consumed by private or industrial consumers.

The life cycle is also influenced by horizontal attributes which have to be carefully assessed in order to allow comparisons among different biofuels: energy balance, emissions, greenhouse gas emissions, other environmental impacts, biofuel costs, and socio-economic impacts.

For example, total costs of biofuels at the filling station include costs for biomass production, biomass transportation, biomass conversion and distribution. Also taxes and profit margins of distributors have to be considered. External costs, like costs for environmental damages, are also important, but they are often neglected.

Environmental criteria for the evaluation of biofuels are mainly energy and greenhouse gas balances. They have to be carefully assessed over the whole life cycle to receive credible results. A general overview of the energy flow and the emissions are shown for all process steps in figure.

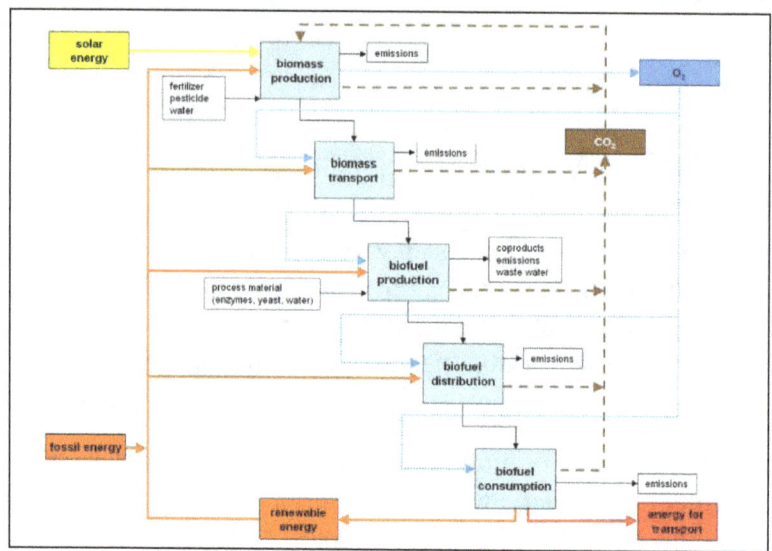

Overview of energy flow and emissions for all
process steps in the life cycle of biofuels.

Finally, biofuels have the potential to create socio-economic benefits. During the life cycle of biofuels, new jobs can be created and agricultural income can be increased. On the other side, labor standards have to be respected and e.g. child labor and slavery has to be avoided.

Energy Balance Methodologies

The energy ratios of biofuels depend on the energy input of the whole life cycle and the energy output for the final fuel. Typically for all biofuels, different steps of the life cycle are characterized by a huge variety which depends on feedstock, agricultural practices, regional feedstock productivity, and process technology. Therefore, the validity of data about biofuel energy balances has to be carefully checked. For example, biofuels from tropical plants have more-favorable energy ratios than biofuels from temperate regions, as tropical crops grow under more favorable climatic conditions. Furthermore, they are often cultivated manually with fewer fossil energy requirements and fewer inputs of fertilizer and pesticides. In contrast, biofuels from temperate regions usually require more energy input. Nevertheless, in recent decades their energy balances have become significantly more efficient.

There are two primary measures for evaluating the energy performance of biofuel production pathways, the energy balance and the energy efficiency:

- The energy balance is the ratio of energy contained in the final biofuel to the energy used by human efforts to produce it. Typically, only fossil fuel inputs are counted in this equation, while biomass inputs, including the biomass feedstock itself, are not counted. A more accurate term for this concept is fossil energy balance, and it is one measure of a biofuel's ability to slow the pace of climate change. The ratio number of the energy balance can exceed one.

- The energy efficiency is the ratio of energy in the biofuel to the amount of energy input, counting all fossil and biomass inputs as well as other renewable energy inputs. This ratio adds an indication of how much biomass energy is lost in the process of converting it to a liquid fuel, and helps to measure more- and less- efficient conversions of biomass to biofuel. The ratio number of the energy efficiency can never exceed one, because some of the energy contained in the feedstock is lost during processing.

The energy balance is a useful metric for biofuel promotion efforts which aim to reduce the use of fossil fuels. For social and ecological reasons it is desirable to reduce fossil energy input in biofuel production. Therefore subsequently mainly the energy balance is quoted.

Fossil transport fuels have energy balances between 0.8 and 0.9. Biofuels significantly contribute to the transportation fuel needs only when these numbers are exceeded. The energy balances of ethanol from wheat, sugar beets and corn are between 1 and 2.5. Ethanol from sugar cane is reported to have an energy balance of approximately 8. The energy balances of lipid derived fuels are between 2.5 and 9. These numbers show that the energy balance of all biofuels is better than that of fossil fuels.

A current problem of evaluating energy balances is the definition of system boundaries. There are debates on whether to include items like the energy required to feed the people processing the feedstock and even the amount of energy a tractor represents. In addition, there is no consensus on what sort of value to give the co-products. In some studies co- products are left on the field to protect the soil from erosion and to add organic matter, while in other studies co-products are used to power the ethanol plant. They do not address the resulting soil erosion which would require energy in the form of fertilizer to replace. In order to get a complete picture about the energy balance of biofuels, at least following variables have to be considered:

- Type of feedstock and agricultural production process.
- Geographical and climate conditions of the producing region.
- Utilized technology for fuel processing.
- Production capacity and scale.
- Sources of process energy.
- Utilization and evaluation of co-products.

Biofuel Emissions

Greenhouse Gas Emissions

One of the major drivers for biofuel developments worldwide is the concern about global climate change which is primarily caused by burning fossil fuels. There is substantial scientific evidence that accelerating global warming is a cause of greenhouse gas (GHG) emissions. One of the main greenhouse gases is carbon dioxide (CO_2). But also nitrous oxide (N_2O), methane (CH_4), several other compounds are greenhouse gases which are even more severe to global warming than CO_2. As their relative potentials for causing global warming differ so much, it has become practice to weight their emissions according to their global warming potentials (GWP) over 100 years and then aggregate them to CO_2 equivalents. GWP is an index for estimating relative global warming contribution due to atmospheric emission of one kg of a particular greenhouse gas compared to emission of one kg of carbon dioxide. GWPs calculated for different time horizons show the effects of atmospheric lifetimes of the different gases. For the assessment of GHG impacts of biofuels mainly CO_2, N_2O and CH_4 are relevant.

Table: Global warming potentials (GWP) of 100 years for several greenhouse gases, relative to carbon dioxide.

Carbon Dioxide (CO_2)	1
Methane (CH_4)	23
Nitrous Oxide (N_2O)	296
HFC-23	12 000
HFC-125	3 400
HFC-134a	1 300
HFC-143a	4 300
HFC-152a	120
HFC-227ea	3 500
HFC-236fa	9 400
Perfluoromethane (CF_4)	5 700
Perfluoroethane (C_2F_6)	11 900
Sulfur Hexafluoride (SF_6)	22 200

As biofuels are produced from biomass, the combustion of these biofuels principally is considered to be CO_2 neutral (this applies only for direct emissions from biofuel combustion). During the combustion process about the same amount of CO_2 is being set free, that has been bound from the atmosphere during photosynthesis and growth of the plants. Therefore the carbon cycle is closed. The major part of combustion engine exhaust streams consists of the components nitrogen, carbon dioxide and water which are non- toxic. However, also greenhouse gases that are directly toxic to human health are emitted. For example the principal transport emissions from the combustion of both fossil and renewable fuels are particulate matter (PM), volatile organic compounds (VOCs) (including hydrocarbons HC), nitrogen oxides (NOx), carbon monoxide (CO) and a variety of unregulated toxic air pollutants. Currently, these emissions (NOx, HC, CO, PM) are regulated for most vehicle types by European emission standards, which are sets of requirements defining the acceptable limits for exhaust emissions of new vehicles sold in EU member states. Apart from direct GHG emissions from burning fuels (which are not accounted in the GHG balance of biofuels as they are renewable), there are significant indirect emissions which are associated with all stages of the biofuel life cycle. For biofuels these emissions are created during cultivation, transport, conversion process and distribution. Thereby emissions from feedstock production are largest in the biofuel life cycle. Nevertheless, it has to be considered, that the life cycle of fossil fuel production produces considerable amounts of emissions, too.

As inputs for the production of biofuels are usually still of fossil origin, the impact on climate of biofuels greatly depends on the fossil energy balance of biofuel production. The combustion and use of fossil sources emit CO_2 which was bound for thousands of years in the earth.

Nevertheless, for the whole assessment of how the production of biofuels influences global warming, not only the fossil energy balance is important, but also other factors have to be included. For example, fertilizing, pesticide use, means of irrigation, and treatment of the soil also play an important role in determining the climate impact of biofuels. One of the most significant factors in terms of climate impact is the use of chemical fertilizers. For their production large inputs of fossil sources usually needed. Fertilizers and in particular nitrogen (N) fertilizers, but also pesticides are manufactured using large input of natural gas. Furthermore, direct emissions from fertilizing may occur. The overall critical point of how biomass production influences climate is the type of feedstock. It determines the energy yield per unit of land, the use of fertilizer, as well as the amount of carbon that can be sequestered in the soil. It also has to be considered what these crops are replacing. If they replace natural grasslands or forests, GHG emissions will likely increase. But if energy crops are planted on unproductive or arid land where conventional crops cannot grow, they have the potential to significantly reduce associated emissions. For example jatropha can thrive on unproductive or arid lands where conventional crops cannot grow. GHG emissions might also be reduced if fuel feedstock replaces annual crops. In this aspect, perennial energy crops are advantageous compared to annual energy crops such as corn or rapeseed. Finally the GHG balance for biofuels could be even more favorable if waste streams, like agricultural and forestry residues, are used as feedstock. Therefore, advanced technologies are necessary, which are not available on a commercial scale today. Finally, the feedstock source also determines the utilization of co- products which has substantial influence on greenhouse gas emissions. Co-products can be used for generating additional renewable energy, for example in combined heat and power plants (CHP).

In conclusion, the emissions during the full life cycle of biofuels, from changes in land use to combustion of fuels, determine their impact on the climate. In modeling this complex calculation, estimates vary widely. Methodologies and calculations of net GHG emissions are highly sensitive to assumptions about system boundaries, key parameter values and how various factors are weighted. Nevertheless, today it is found in nearly all studies that GHG emissions from first generation ethanol and biodiesel are significantly reduced, relative to fossil fuels. There exists broad agreement that the use of biofuels, made with today's technologies, can result in significant net reductions in carbon emissions.

An appropriate method in determining the environmental impacts of GHG emissions is the so called Well-to-Wheel analysis (WTW). This approach can be divided into a Well-to-Tank analysis (WTT) which includes cultivation and conversion process and Tank-to-Wheel analysis (TTW) which analyses the use of biofuels itself and includes vehicle and engine performance. Currently GHG calculation tools for biofuels from different feedstocks are being developed by the institutes Imperial College London, UK, Senter Noven, The Netherlands, and ifeu Institute, Germany.

Sustainability of Biofuels

The growing market for biofuels stimulated discussion about sustainability of biofuels and pushed the call for sustainability standards. This includes environmental and social aspects.

Within the discussion about negative social impacts, compliance with laborers rights, prevention of child labor, and implementation of minimum working conditions are among the most important issues. This includes also gender issues, land use rights, food versus fuels, health and safety, quality of life, and education.

The most important issues in the discussion about negative environmental impacts are, apart from GHG savings, destruction of rain forests and wetlands, loss of biodiversity, water pollution, acidification, eutrophication, and impact on ground source water. These environmental impacts are mainly associated with agricultural practices and feedstock production. But also impacts of biomass transport, biofuel production, distribution, and consumption have to be considered in the life cycle analysis.

The size of negative impacts depends on different parameters, and especially on the practices of feedstock producers. But if feedstock production is done in a sustainable manner, environmental impacts can also be positive. For example dedicated perennial energy crops can prevent soil erosion and the practices of double cropping, crop mixing and the plantation of second generation feedstock can even enhance biodiversity.

Because of the large variety of feedstock types and its production processes, social and environmental impacts of biofuel production can not be evaluated in general. They depend on local conditions and on the amount of land necessary to cultivate biofuel feedstock. Thus, for each case social and environmental impacts have to be evaluated separately.

Economy of Biofuels

In the whole life cycle of biofuels, the relatively high production costs still remain a critical barrier to commercial development, although continuing improvements are achieved. Nevertheless, technologies for pure plant oil and biodiesel production from oilseed crops are already fairly mature.

The competitiveness of biofuels will increase as prices for crude oil and other fossil sources increase and overstep a critical threshold. Today biofuel competitiveness still largely depends on the national legislative frameworks and subsidies in EU member states. Subsidies can be both agricultural aids and market incentives for the biofuel itself. Also tax exemptions have considerable impacts on end-user costs for biofuels.

For first generation biofuels, the feedstock is a major component of overall costs. As crop prices are highly volatile, the overall production costs of biofuels vary. The production scale of biofuels has significant impact on cost. It is more important for ethanol processing than for production of pure plant oil and biodiesel. This advantage for lipid derived fuels is especially important for small scale agricultural producers and SME's.

Thus, for example in Germany, biodiesel is currently mainly produced by small scale producers at relatively low process costs.

Generally biofuels are expected to have large socio-economic impacts, especially for local actors. Biofuel production opens new market opportunities for agricultural products and thus new income options for farmers. In the future agriculture will not only play a role in food production, but also in energy provision. The increased feedstock production is expected to strongly contribute to the multi-functionality of the agricultural sector. Nevertheless it is difficult to assess the real dimension of additional employment and impact on local economy in the biomass sector. On EU level no detailed study has been conducted on this topic.

Second generation fuels are not yet produced on commercial scale. Due to high production costs, they are not competitive at the moment, but as technology improves, they may become an important role in biofuel provision. The great advantage of these fuels is the vast range of feedstock that can be used for biofuel production, as well as the reduced feedstock (e.g cellulose crops) costs.

By using a holistic approach, biofuels offer large economic advantages over fossil fuels, but direct cost comparisons are difficult. Negative externalities associated with fossil fuels tend to be poorly quantified, such as military expenditures and costs for health and environment. However, biofuels have the potential to generate many positive externalities, such as reduced greenhouse gas emissions, decreased air pollution, and job creation. Additionally biofuels decrease dependency from crude oil imports. Consequently biofuels are a more socially and environmentally desirable liquid fuel, a fact that is often neglected in direct-cost calculations. Therefore biofuels often seem uncompetitive although a biofuel market may actually provide long-term economic benefits when comparing environmental and social costs.

Consideration of Co-products

During biofuel production large quantities of co-products are received. The utilization of these co-products contributes to an increase of energy efficiency of the whole process by subsidizing other products like mineral fertilizer made from fossil fuel. Co-products also reduce GHG emissions and constitute an additional economical value. But it is difficult to value and predict the benefits of co-products. Especially when an increase of ethanol production enhances the supply of co-products, reactions of the market can hardly be foreseen.

The different methods for consideration of co-products in the process of ethanol biofuel production are:

- Target product orientated method: This very simple method is used in many old studies. Thereby the whole energy input is charged to the target product (e.g. ethanol), co-products are not considered. The energetic and economic use of co- products is neglected and no holistic approach is implemented. This conservative method represents the worst-case scenario, but uncertainties and vagueness are eliminated.

- Allocation method: Environmental effects are determined and allocated separately to the co-product and to the target product.

- Credit method: The energy content of co-products is credited to the target product. Precondition is the real use of the co-product. Additional energy inputs for the preparation of the co-products have to be considered.

- Substitution method: As co-products substitute similar products, environmental effects can be avoided. These effects are credited to the target product.

In the fermentation of sugar and starch-bearing plants, co-products are produced in large quantities. They can be used as fodder, fertilizer, heat fuel, industrial raw material or as substrate for biogas plants. An excellent example how co- products from ethanol production can be used is the bagasse, the fibrous residuals of sugar cane after pressing. In Brazil, bagasse is burned and the heat is used for the distillation process and for electricity generation.

Similar, large quantities of co-products are received from the production of lipid derived fuels, such as biodiesel and PPO. For instance, press cake from rapeseed oil extraction is a high value and protein rich fodder. In biodiesel production glycerin is a valuable co- product for industrial purposes.

Co-products: bagasse from sugarcane (left) and rape seed cake (right).

TYPES OF BIOFUELS

Algal Fuel

Algal biofuel is an alternative to fossil fuel, which is generated by specific algae species from carbon dioxide.

These algae species are primarily unicellular or diatom microalgae that produce high carbohydrate compositions suitable for ethanol production, high lipid compositions suitable for biodiesel production or high hydrocarbon compositions that are suitable for producing renewable distillates.

Increase in fuel costs and consumption, and depletion of natural fuel resources have created a demand for research into alternative forms of fuels in the last decade. Several companies and government agencies are funding research to try and make algae fuel production commercially viable.

The optimum selection of the algal species for biofuel production is based on the ability to sustain the culture, growth rate of the species, the biomass specific contents of proteins, carbohydrates, and lipids, and the overall supporting photosynthesis environment.

Fuels from Algae

The lipid (oily) part of the algae biomass can be extracted and converted into biodiesel by a process similar to that used for any other vegetable oil.

Butanol can be made from algae or diatoms using a solar-powered biorefinery. This fuel was found to have an energy density 10% less than gasoline, and greater than that of either methanol or ethanol.

The green waste left over from the algae oil extraction can be used to produce butanol. Additionally, it was found that macroalgae can be fermented by Clostridria to form butanol and other solvents.

Biogasoline produced from algae biomass can be used in internal combustion engines. Methane, which is the chief component of natural gas, can be produced from algae using several methods - pyrolysis, gasification or anaerobic digestion.

Algae can also be used to produce green diesel, also known as renewable diesel through a hydrocracking refinery process that breaks down molecules into shorter hydrocarbon chains used in diesel engines.

Algae Cultivation

Algae can produce up to 300 times more oil per unit area than conventional crops such as palms, soybeans, rapeseed or jatroba. The following three primary ways to grow algae for biofuel production have been identified:

Open Pond System

The open pond system is one of the easiest methods for the cultivation of algae with high-oil content. In this method, algae are grown in open ponds under very warm and sunny environments.

Although it is the simplest form of algae production, it also has some major drawbacks. Open systems using a monoculture are also vulnerable to viral infection. In order to enhance algae production using this method, water temperature needs to be controlled.

Closed-loop System

The closed-loop system was adapted to produce algae more quickly and efficiently than the open pond system. In this method, algae are placed in clear, plastic bags to allow them to be exposed to sunlight.

These bags are stacked high and protected from external elements using a cover. The clear plastic bag provides enough exposure to sunlight to increase the rate of algae production.

The greater the algae production, the greater the amount of oil will be extracted. Unlike the open pond method, this method prevents algal contamination.

Photobioreactors

Most of the companies that use algae as a source of biofuels employ borosilicate glass tubes known as bioreactors that are exposed to sunlight. Within these tubes, the algae can be grown at maximum levels, even to the point they can be harvested every day.

This method results in a very high output of algae and oil for producing biofuels. However, running a photobioreactor is more expensive and difficult than using the open pond system, but may provide a high level of control.

Benefits of Algal Fuel

The key benefits of algae and algal fuels are listed below:

- Algae require much less land to grow when compared to other traditional row crops, such as corn. In addition, algae can be grown on non-arable, nutrient-poor land that does not support conventional agriculture.

- Algae farms for producing biofuel can thrive without petroleum-based fertilizers, fresh water for irrigation or arable land.

- Algae can be grown rapidly at large scale and generate up to 50 times more oil per acre than other row crops like soybeans and corn.

- Algae biofuels help reduce the country's energy dependence.

- Algae use photosynthesis to capture sunlight energy for producing carbohydrates and oxygen thereby creating a natural biomass oil product.

- Algae can grow in seawater as well as high-saline water. Several species of algae can also grow in wastewater from treatment plants and water-containing phosphates, nitrates, and other contaminants.

- Algae fuels are biodegradable and non-toxic as they do not contain sulfur.

- Unlike fossil fuels, harvested algae release CO_2 when burnt, but it is absorbed by new growing algae.

Methanol Fuel

Methanol fuel is an alternative biofuel for internal combustion and other engines, either in combination with gasoline or independently. Methanol is less expensive to produce sustainably than ethanol fuel, although it is generally more toxic and has lower energy density. For optimizing engine performance and fuel availability, however, a blend of ethanol, methanol and petroleum is likely to be preferable to using any of these alone. Methanol may be made from hydrocarbon or renewable resources, in particular natural gas and biomass respectively. It can also be synthesized from CO_2 (carbon dioxide) and hydrogen. Methanol fuel is currently used by racing cars in many countries but has not seen widespread use otherwise.

Production

Historically, methanol was first produced by destructive distillation (pyrolysis) of wood, resulting in its common English name of wood alcohol.

At present, methanol is usually produced using methane (the chief constituent of natural gas) as a raw material. In China, methanol is made for fuel from coal.

"Biomethanol" may be produced by gasification of organic materials to synthesis gas followed by conventional methanol synthesis. This route can offer methanol production from biomass at efficiencies up to 75%. Widespread production by this route has a proposed potential to offer methanol fuel at a low cost and with benefits to the environment. These production methods, however, are not suitable for small-scale production.

Recently, methanol fuel has been produced using renewable energy and carbon dioxide as a feedstock. Carbon Recycling International, an Icelandic-American company, completed the first commercial scale renewable methanol plant in 2011.

Major Fuel use

During the OPEC 1973 oil crisis, Reed and Lerner proposed methanol from coal as a proven fuel with well-established manufacturing technology and sufficient resources to replace gasoline. Hagen reviewed prospects for synthesizing methanol from fossil and renewable resources, its use as a fuel, economics, and hazards. Then in 1986, the Swedish Motor Fuel Technology Co. (SBAD) extensively reviewed the use of alcohols and alcohol blends as motor fuels. It reviewed the potential for methanol production from natural gas, very heavy oils, bituminous shales, coals, peat and biomass. In 2005, 2006 Nobel prize winner George A. Olah, G. K. Surya Prakash and Alain Goeppert advocated an entire methanol economy based on energy storage in synthetically produced methanol. The Methanol Institute, the methanol trade industry organization, posts reports and presentations on methanol. Director Gregory Dolan presented the 2008 global methanol fuel industry in China.

On January 26, 2011, the European Union's Directorate-General for Competition approved the Swedish Energy Agency's award of 500 million Swedish kronor (approx. €56M as at January 2011) toward the construction of a 3 billion Swedish kronor (approx. €335M) industrial scale experimental development biofuels plant for production of Biomethanol and BioDME at the Domsjö Fabriker biorefinery complex in Örnsköldsvik, Sweden, using Chemrec's black liquor gasification technology.

Uses

Internal Combustion Engine Fuel

Both methanol and ethanol burn at lower temperatures than gasoline, and both are less volatile, making engine starting in cold weather more difficult. Using methanol as a fuel in spark-ignition engines can offer an increased thermal efficiency and increased power output (as compared to gasoline) due to its high octane rating (114) and high heat of vaporization. However, its low energy content of 19.7 MJ/kg and stoichiometric air-to-fuel ratio of 6.42:1 mean that fuel consumption (on volume or mass bases) will be higher than hydrocarbon fuels. The extra water produced also makes the charge rather wet (similar to hydrogen/oxygen combustion engines) and with the formation of acidic products during combustion, the wearing of valves, valve seats and cylinder might be higher than with hydrocarbon burning. Certain additives may be added to the fuel in order to neutralize these acids.

Methanol, like ethanol, contains soluble and insoluble contaminants. These soluble contaminants, halide ions such as chloride ions, have a large effect on the corrosivity of alcohol fuels. Halide ions increase corrosion in two ways; they chemically attack passivating oxide films on several metals causing pitting corrosion, and they increase the conductivity of the fuel. Increased electrical conductivity promotes electric, galvanic, and ordinary corrosion in the fuel system. Soluble contaminants, such as aluminum hydroxide, itself a product of corrosion by halide ions, clog the fuel system over time.

Methanol is (in automotive terms) hygroscopic, meaning it will absorb water vapor directly from the atmosphere. Because absorbed water dilutes the fuel value of the methanol (although it suppresses engine knock), and may cause phase separation of methanol-gasoline blends, containers of methanol fuels must be kept tightly sealed.

Compared to gasoline, methanol is more tolerant to exhaust gas recirculation (EGR), which improves fuel efficiency of the internal combustion engines utilizing Otto cycle and spark ignition.

An acid, albeit weak, methanol attacks the oxide coating that normally protects the aluminium from corrosion:

$$6\,CH_3OH + Al_2O_3 \rightarrow 2\,Al(OCH_3)_3 + 3\,H_2O$$

The resulting methoxide salts are soluble in methanol, resulting in a clean aluminium surface, which is readily oxidized by dissolved oxygen. Also, the methanol can act as an oxidizer:

$$6\,CH_3OH + 2\,Al \rightarrow 2\,Al(OCH_3)_3 + 3\,H_2$$

This reciprocal process effectively fuels corrosion until either the metal is eaten away or the concentration of CH_3OH is negligible. Methanol's corrosivity has been addressed with methanol-compatible materials and fuel additives that serve as corrosion inhibitors.

Organic methanol, produced from wood or other organic materials (bioalcohol), has been suggested as a renewable alternative to petroleum-based hydrocarbons. Low levels of methanol can be used in existing vehicles with the addition of cosolvents and corrosion inhibitors.

Racing

Pure methanol is required by rule to be used in Champcars, Monster Trucks, USAC sprint cars (as well as midgets, modifieds, *etc.*), and other dirt track series, such as World of Outlaws, and Motorcycle Speedway, mainly because, in the event of an accident, methanol does not produce an opaque cloud of smoke. Since the late 1940s, Methanol is also used as the primary fuel ingredient in the powerplants for radio control, control line and free flight model aircraft , cars and trucks; such engines use a platinum filament glow plug that ignites the methanol vapor through a catalytic reaction. Drag racers, mud racers, and heavily modified tractor pullers also use methanol as the primary fuel source. Methanol is required with a supercharged engine in a Top Alcohol Dragster and, until the end of the 2006 season, all vehicles in the Indianapolis 500 had to run on methanol. As a fuel for mud racers, methanol mixed with gasoline and nitrous oxide produces more power than gasoline and nitrous oxide alone.

Beginning in 1965, pure methanol was used widespread in USAC Indy car competition, which at the time included the Indianapolis 500.

Safety was the predominant influence for the adoption of methanol fuel in the United States open-wheel racing categories. Unlike petroleum fires, methanol fires can be extinguished with plain water. A methanol-based fire burns invisibly, unlike gasoline, which burns with a visible flame. If a fire occurs on the track, there is no flame or smoke to obstruct the view of fast approaching drivers, but this can also delay visual detection of the fire and the initiation of fire suppression. A seven-car crash on the second lap of the 1964 Indianapolis 500 resulted in USAC's decision to encourage, and later mandate, the use of methanol. Eddie Sachs and Dave MacDonald died in the crash when their gasoline-fueled cars exploded. The gasoline-triggered fire created a dangerous cloud of thick black smoke that completely blocked the view of the track for oncoming cars. Johnny Rutherford, one of the other drivers involved, drove a methanol-fueled car, which also leaked following the crash. While this car burned from the impact of the first fireball, it formed a much smaller inferno than the gasoline cars, and one that burned invisibly. That testimony, and pressure from *The Indianapolis Star* writer George Moore, led to the switch to alcohol fuel in 1965.

Methanol was used by the CART circuit during its entire campaign. It is also used by many-short track organizations, especially midget, sprint cars and speedway bikes. Pure methanol was used by the IRL from 1996-2006.

In 2006, in partnership with the ethanol industry, the IRL used a mixture of 10% ethanol and 90% methanol as its fuel. Starting in 2007, the IRL switched to "pure" ethanol, E100.

Methanol fuel is also used extensively in drag racing, primarily in the Top Alcohol category, while between 10% and 20% methanol may be used in Top Fuel classes in addition to Nitromethane.

Formula One racing continues to use gasoline as its fuel, but in prewar grand prix racing methanol was often used in the fuel. Methanol is also used in Monster Truck racing.

Fuel for Model Engines

The earliest model engines for free-flight model aircraft flown before the end of World War II used a 3:1 mix of white gas and heavy viscosity motor oil for the two-stroke spark ignition engines used for the hobby at that time. By 1948, the new glow plug-ignition model engines began to take over the market, requiring the use of methanol fuel to react in a catalytic reaction with the coiled platinum filament in a glow plug for the engine to run, usually using a castor oil-based lubricant contained in the fuel mix at about a 4:1 ratio. The glow-ignition variety of model engine, because it no longer required an onboard battery, ignition coil, ignition points and capacitor that a spark ignition model engine required, saved valuable weight and allowed model aircraft to have better flight performance. In their traditionally popular two-stroke and increasingly popular four-stroke forms, currently produced single cylinder methanol-fueled glow engines are the usual choice for radio controlled aircraft for recreational use, for engine sizes that can range from 0.8 cm³ (0.049 cu.in.) to as large as 25 to 32 cm³ (1.5-2.0 cu.in) displacement, and significantly larger displacements for twin and multi-cylinder opposed-cylinder and radial configuration model aircraft engines, many of which are of four-stroke configuration. Most methanol-fueled model engines, especially those made outside North America, can easily be run on so-called *FAI*-specification methanol fuel. Such fuel mixtures can be required by the FAI for certain events in so-called FAI "Class F" international competition, that forbid the use of nitromethane as a glow engine fuel component. In contrast, firms in North America that make methanol-fueled model engines, or who are based outside that continent and have a major market in

North America for such miniature powerplants, tend to produce engines that can and often do run best with a certain percentage of nitromethane in the fuel, which when used can be as little as 5% to 10% of volume, and can be as much as 25 to 30% of the total fuel volume.

Cooking

Methanol is used as a cooking fuel in China and its use in India is growing. Its stove and canister need no regulators or pipes.

Toxicity

Methanol occurs naturally in the human body and in some fruits, but is poisonous in high concentration. Ingestion of 10 ml can cause blindness and 60-100 ml can be fatal if the condition is untreated. Like many volatile chemicals, methanol does not have to be swallowed to be dangerous since the liquid can be absorbed through the skin, and the vapors through the lungs. Methanol fuel is much safer when blended with ethanol, even at relatively low ethanol percentages.

US maximum allowed exposure in air (40 h/week) is 1900 mg/m^3 for ethanol, 900 mg/m^3 for gasoline, and 1260 mg/m^3 for methanol. However, it is much less volatile than gasoline, and therefore has lower evaporative emissions, producing a lower exposure risk for an equivalent spill. While methanol offers somewhat different toxicity exposure pathways, the effective toxicity is no worse than those of benzene or gasoline, and methanol poisoning is far easier to treat successfully. One substantial concern is that methanol poisoning generally must be treated while it is still asymptomatic for full recovery.

Inhalation risk is mitigated by a characteristic pungent odor. At concentrations greater than 2,000 ppm (0.2%) it is generally quite noticeable, however lower concentrations may remain undetected while still being potentially toxic over longer exposures, and may still present a fire/explosion hazard. Again, this is similar to gasoline and ethanol; standard safety protocols exist for methanol and are very similar to those for gasoline and ethanol.

Use of methanol fuel reduces the exhaust emissions of certain hydrocarbon-related toxins such as benzene and 1,3 butadiene, and dramatically reduces long term groundwater pollution caused by fuel spills. Unlike benzene-family fuels, methanol will rapidly and non-toxically biodegrade with no long-term harm to the environment as long as it is sufficiently diluted.

Fire Safety

Methanol is far more difficult to ignite than gasoline and burns about 60% slower. A methanol fire releases energy at around 20% of the rate of a gasoline fire, resulting in a much cooler flame. This results in a much less dangerous fire that is easier to contain with proper protocols. Unlike gasoline fires, water is acceptable and even preferred as a fire suppressant for methanol fires, since this both cools the fire and rapidly dilutes the fuel below the concentration where it will maintain self-flammability. These facts mean that, as a vehicle fuel, methanol has great safety advantages over gasoline. Ethanol shares many of these same advantages.

Since methanol vapor is heavier than air, it will linger close to the ground or in a pit unless there is good ventilation, and if the concentration of methanol is above 6.7% in air it can be lit by a spark and will explode above 54 F / 62 C. Once ablaze, an undiluted methanol fire gives off very little visible light, making it potentially very hard to see the fire or even estimate its size in bright daylight, although in the vast majority of cases, existing pollutants or flammables in the fire (such as tires or asphalt) will color and enhance the visibility of the fire. Ethanol, natural gas, hydrogen, and other existing fuels offer similar fire-safety challenges, and standard safety and firefighting protocols exist for all such fuels.

Post-accident environmental damage mitigation is facilitated by the fact that low-concentration methanol is biodegradable, of low toxicity, and non-persistent in the environment. Post-fire clean-up often merely requires large additional amounts of water to dilute the spilled methanol followed by vacuuming or absorption recovery of the fluid. Any methanol that unavoidably escapes into the environment will have little long-term impact, and with sufficient dilution will rapidly biodegrade with little to no environmental damage due to toxicity. A methanol spill that combines with an existing gasoline spill can cause the mixed methanol/gasoline spill to persist about 30% to 35% longer than the gasoline alone would have done.

Use

United States

The State of California ran an experimental program from 1980 to 1990 that allowed anyone to convert a gasoline vehicle to 85% methanol with 15% additives of choice. Over 500 vehicles were converted to high compression and dedicated use of the 85/15 methanol and ethanol.

In 1982 the big three were each given $5,000,000 for design and contracts for 5,000 vehicles to be bought by the State. It was an early use of low-compression flexible-fuel vehicles.

In 2005, California's Governor, Arnold Schwarzenegger, stopped the use of methanol to join the expanding use of ethanol driven by producers of corn. In 2007 ethanol was priced at 3 to 4 dollars per gallon (0.8 to 1.05 dollars per liter) at the pump, while methanol made from natural gas remains at 47 cents per gallon (12.5 cents per liter) in bulk, not at the pump.

Presently there are no operating gas stations in California supplying methanol in their pumps. Rep. Eliot Engel [D-NY17] has introduced "An Open Fuel Standard" Act in Congress: "To require automobile manufacturers to ensure that not less than 80 percent of the automobiles manufactured or sold in the United States by each such manufacturer to operate on fuel mixtures containing 85 percent ethanol, 85 percent methanol, or biodiesel."

European Union

The amended Fuel Quality Directive adopted in 2009 allows up to 3% v/v blend-in of methanol in petrol.

Brazil

A drive to add an appreciable percentage of methanol to gasoline got very close to implementation in Brazil, following a pilot test set up by a group of scientists involving blending gasoline with methanol between 1989 and 1992. The larger-scale pilot experiment that was to be conducted in

São Paulo was vetoed at the last minute by the city's mayor, out of concern for the health of gas station workers, who would not be expected to follow safety precautions. As of 2006, the idea has not resurfaced.

India

Niti Aayog, The planning commission of India on 3rd August 2018 announced that if feasible, passenger vehicles will run on 15% Methanol mixed fuel. At present, vehicles in India use up to 10% ethanol-blended fuel. If approved by the government it will cut monthly fuel costs by 10%. In India ethanol costs Rs 42 a litre, while the price of methanol has been estimated at less than Rs 20 a litre.

Ethanol Fuel

Ethanol, which is sometimes known as ethyl alcohol, is a kind of alcohol derived from corn, sugarcane, and grain or indirectly from paper waste. It's also the main type of alcohol in most alcoholic beverages obtained as a result of fermentation of a mash of grains (gin, vodka, and whiskey) or sugarcane (rums). It's also a source of fuel commonly blended with gasoline to oxygenate the fuel at the gas pump. Ethanol fuel can also be used on its own to power vehicles.

Ethanol is more common in our lives than you may think. After all, any alcoholic beverage you can drink comprises of Ethanol. It is known by many different names such as Ethyl alcohol, pure alcohol and grain alcohol. It is regarded as an alternative form of fuel that has gained much popularity for a number of reasons.

The most common use of Ethanol fuel is by blending it with gasoline. Doing so creates a mix that releases fewer emissions into the environment and is considered cleaner in nature. It also keeps the car in a better shape by increasing the octane rating of the fuel. All in all, it is accepted by the people, governments and car companies for the many benefits it provides.

Ethanol does not occur naturally in any eco-system. It is produced through the processes of fermentation and distillation. While the energy based use of Ethanol fuel is new, it has been part of our lives for a very long time. Fermenting sugar creates Ethanol – knowledge used by our forefathers. These days, it comes from crops and plants that are rich in sugar or have the ability to be converted into cellulose and starch. Sugarcane, barley, sugar beets, wheat and corn are commonly used for production.

Transformation of Ethanol into Fuel

The process starts by grinding up the crops or plants meant for production. After this, the ground up substance is refined to get sugar, cellulose or starch. Sugar from plant material is converted into ethanol and carbon dioxide by fermentation. Yeast is normally added to speed up the fermentation process (just the same way alcoholic beverages are produced). Once the ethanol is distilled and purified, it is ready for use. Having a four-step process like this allows the production to be comparatively cost-effective, which is one big reason for the use of Ethanol fuel in our current economy.

To make ethanol fuel from sugarcane, you need to squeeze out the juice from the sugarcane, ferment and then distil it. Compared to the traditional unleaded gasoline, ethanol is a clean-burning, particulate-free fuel source. When burnt with oxygen, the end product is carbon dioxide and water.

Ethanol fuel is not a trend that has come in recently and will die out soon. Governments and automobile manufacturers have recognized the benefits of using it and are working towards integrating it into everyday use. A number of vehicles now come designed with engines that can work with the standard gasoline-ethanol blend. All of this because there are many known benefits of using this form of fuel.

Advantages of Ethanol Fuel

1. Ethanol Fuel is Cost Effective Compared to other Biofuels

    Ethanol fuel is the least expensive energy source since virtually every country has the capability to produce it. Corn, sugar cane or grain grows in almost every country which makes the production economical compared to fossil fuels. Fossils fuels can play against the economy of most countries, especially, developing countries that have no capacity to explore them. It, thus, makes sense for these growing economies to dwell on the production of ethanol fuel to dial back on the dependence of fossil fuel in order to save revenue.

2. Ecologically Effective

    One striking advantage of ethanol over other fuel sources is that it does not cause pollution to the environment. Using ethanol fuel to power automobiles results in significantly low levels of toxins in the environment. On numerous occasions, ethanol is converted to fuel by blending with gasoline. Specifically, ethanol to gasoline ratio of 85:15. The little composition of gasoline acts as an igniter, while ethanol takes up the rest of the tasks. This ratio of ethanol to gasoline minimizes the emission of greenhouse gases to the environment since it burns cleanly compared to pure gasoline.

3. Minimizes global warming

    Global warming is caused by relentless emission of dangerous greenhouse gases emanation from use of fossil fuels (oil, natural gas, and coal). The effects of global warming are catastrophic including changes in weather patterns, rising sea levels, and excessive heat. Combustion of ethanol fuel only releases carbon dioxide and water. The carbon dioxide released is ineffective regarding environment degradation.

4. Easily accessible

    Since ethanol is a biofuel, it is easily accessible to virtually everyone. Biofuel means energy derived from plants like sugarcane, grains, and corn. All tropical climates support growth of sugarcane. Grain and corn grow in every country. In fact, corn is a staple food in most countries in Africa.

5. Minimizes dependence on fossil fuels

    Harnessing of fuel from corn or biomass is an economical way to sustain any economy and prevent it from over-reliance on importation of fossil fuels like oil, and gas. Embracing ethanol fuel can save a country a lot of money that can be plowed back into the economy. Since ethanol is domestically produced, from domestically grown crops, it help reduce dependance on foreign oil and greenhouse gas emissions. If we could run our vehicles on 100% ethanol, the difference would be noticeable.

6. Contributes to creation of employment to the country

    When the use of ethanol fuel increases, it means more plantations of sugarcane, corn, and grains. It also means more ethanol fuel processing plants and that translates to job opportunities. Ethanol can also be branched out to produce alcoholic beverages leading to creation of job opportunities in the hospitality industry.

7. Opens up untapped agricultural sector

    The fact that ethanol fuel production relies mainly on agricultural produce, individuals will be shoved into the untapped agricultural sector, and this will uplift a countries economy. This act will guarantee ethanol fuel availability for many years. The need for increased production of corn and grains has set the farming industry booming.

8. Ethanol fuel is a source of hydrogen

    Although ethanol fuel is not perfect, researchers are working around the clock to beef up its efficiency to make it a reliable energy source by getting rid of its disadvantages. One disadvantage of ethanol fuel is that it has been reported to cause engine burns and corrosion. To be able to utilize it in a more productive way, researchers are looking to convert it into hydrogen form, which should uplift it as a formidable alternative source of fuel.

9. Variety of sources of raw material

    Although corn and sugarcane are the chief raw material for producing ethanol fuel, pretty much every crop or plant containing starch and sugar can be used.

10. Ethanol is classified as a renewable energy source

    It's classified as a renewable resource because it's mainly as a consequence of conversion of energy from the sun into useful energy. The production of ethanol begins with the photosynthesis process, which enables sugarcane to thrive and later be processed into ethanol fuel.

Disadvantages of Ethanol Fuel

1. Requires large piece of land

    Ethanol is produced from corn, sugarcane, and grains. All these are crops that need to be grown in farms. For ethanol to meet the growing demand, it must be produced in large scale. This, essential, means that these very crops will have to be grown in large scale, which requires vast acres of land. The problem is that not everyone has that kind of land, so the only option is renting or leasing, which might add expenses to the budget. This aspect could also lead to destruction of natural habitats for most plants and animals.

2. Distillation process is not good for environment

    The process of distilling fermented corn or grain takes a long time and involves a lot of heat expenditure. The source of heat for distillation is mostly fossil fuel, and fossil fuels emit a lot of greenhouse gas, which is detrimental to the environment.

3. Spike in food prices

    The chief ingredient in making ethanol is corn. If the demand for ethanol fuel skyrockets, the price of corn would also shoot up, and that would affect the cost of ethanol production. Other users of corn other than for fuel will also suffer, for example, those utilizing corn as an animal feed. Also, the lucrative prices of ethanol fuel could trigger most farmers to abandon food crops for ethanol production, which might also lead to an increase in food prices.

4. Water attraction

    Pure ethanol has high affinity for water, and it's able to absorb any trace around it or from the atmosphere. This fact is also true for those blends of gasoline and ethanol used to power vehicles. The fact that ethanol has high water attraction capabilities means that it's difficult to obtain it in its purest form since there will somehow be a trace of water. In fact, manufacturers normally indicate 99.8% pure ethanol. This is especially dangerous for marine users than regular road car users.

    When water finds way into a storage or fuel tank, it goes to the bottom of tank since water is denser than fuel. This will lead to a plethora of small and big engine problems for your vehicle. The water attraction property of ethanol is the reason why it's transported by railroad or auto transport.

5. Difficult to vaporize

    Pure ethanol is hard to vaporize. This makes starting a car in cold conditions almost difficult, which is why a number of vehicle owners make a point to retain a little petrol, for instance, E85 cars that use 15% petroleum and 85% ethanol.

    A common blend used these days is E85 i.e. 85% Ethanol and 15% gasoline. The mileage provided by this blend is lesser than that of pure gasoline or the E10 (10% Ethanol) blend. However, the benefit of using the E85 blend is that the oil remains clean for a longer time, there is lesser stress on the engine and the overall engine maintenance reduces. The cost of lower mileage gets covered up thanks to these small benefits. Not to mention, the overall reduction of your carbon footprint, which is the one benefit from the use of Ethanol fuel that everybody should aspire for.

Butanol Fuel

Butanol may be used as a fuel in an internal combustion engine. Because its longer hydrocarbon chain causes it to be fairly non-polar, it is more similar to gasoline than it is to ethanol. Butanol is a drop-in fuel and thus works in vehicles designed for use with gasoline without modification. It has a four link hydrocarbon chain. It can be produced from biomass (as "biobutanol") as well as fossil fuels (as "petrobutanol"), but biobutanol and petrobutanol have the same chemical properties.

Production of Biobutanol

Butanol from biomass is called biobutanol. It can be used in unmodified gasoline engines. High cost of raw material is considered as one of the main barriers against commercial butanol fermentation. Using inexpensive and abundant feedstocks, e.g., corn stover, can enhance the process economic viability.

Technologies

Biobutanol can be produced by fermentation of biomass by the A.B.E. process. The process uses the bacterium Clostridium acetobutylicum, also known as the Weizmann organism, or Clostridium beijerinckii. It was Chaim Weizmann who first used C. acetobutylicum for the production of acetone from starch (with the main use of acetone being the making of Cordite) in 1916. The butanol was a by-product of fermentation (twice as much butanol was produced). The process also creates a recoverable amount of H2 and a number of other by-products: acetic, lactic and propionic acids, isopropanol and ethanol.

Biobutanol can also be made using *Ralstonia eutropha* H16. This process requires the use of an electro-bioreactor and the input of carbon dioxide and electricity.

The difference from ethanol production is primarily in the fermentation of the feedstock and minor changes in distillation. The feedstocks are the same as for ethanol: energy crops such as sugar beets, sugar cane, corn grain, wheat and cassava, prospective non-food energy crops such as switchgrass and even guayule in North America, as well as agricultural byproducts such as bagasse, straw and corn stalks. According to DuPont, existing bioethanol plants can cost-effectively be retrofitted to biobutanol production.

Additionally, butanol production from biomass and agricultural byproducts could be more efficient (i.e. unit engine motive power delivered per unit solar energy consumed) than ethanol or methanol production.

Algae Butanol

Biobutanol can be made entirely with solar energy and nutrients, from algae (called Solalgal Fuel) or diatoms. Current yield is low.

Research

Although biofuel demand has risen to over one billion liters (about 260 million US gallons) yearly, fermentation remains a largely inefficient method of butanol production. Under normal living conditions, *Clostridium* bacterial communities have a low yield of butanol per gram of glucose. Obtaining higher yields of butanol involves manipulation of the metabolic networks within bacteria to prioritize the synthesis of the biofuel. Metabolic engineering and genetic engineering tools allow scientists to alter the states of reactions occurring in the organism, utilizing advanced techniques to create a bacterial strain capable of high butanol yield. Optimization can also be accomplished by the transfer of specific genetic information to other uni-cellular species, capitalizing on the traits of multiple organisms to achieve the highest rate of alcohol production.

Using Alternate Carbon Sources

One promising development in biobutanol production technology was discovered in the late summer of 2011—Tulane University's alternative fuel research scientists discovered a strain of Clostridium, called "TU-103", that can convert nearly any form of cellulose into butanol, and is the only known strain of Clostridium-genus bacteria that can do so in the presence of oxygen. The university's researchers have stated that the source of the "TU-103" Clostridium bacteria strain was most likely from the solid waste from one of the plains zebra at New Orleans' Audubon Zoo.

Metabolic engineering can be used to allow an organism to use a cheaper substrate such as glycerol instead of glucose. Because fermentation processes require glucose derived from foods, butanol production can negatively impact food supply .Glycerol is a good alternative source for butanol production. While glucose sources are valuable and limited, glycerol is abundant and has a low market price because it is a waste product of biodiesel production. Butanol production from glycerol is economically viable using metabolic pathways that exist in *Clostridium pasteurianum* bacterium.

A combination of succinate and ethanol can be fermented to produce butyrate (a precursor to butanol fuel) by utilizing the metabolic pathways present in a gram-positive anaerobic bacterium Clostridium kluyveri. Succinate is an intermediate of the TCA cycle, which metabolizes glucose. Anaerobic bacteria such as Clostridium acetobutylicum and Clostridium saccharobutylicum also contain these pathways. Succinate is first activated and then reduced by a two-step reaction to give 4-hydroxybutyrate, which is then metabolized further to crotonyl-coenzyme A (CoA) . Crotonyl-CoA is then converted to butyrate. The genes corresponding to these butanol production pathways from Clostridium were cloned to E. coli.

In 2012 researchers developed a method for storing electrical energy as chemical energy in higher alcohols (including butanol). These alcohols can then be used as liquid transportation fuels. The team led by James C. Liao genetically engineered lithoautotrophic microorganism known as Ralstonia Eutropha H16 to produce isobutanol and 3-methyl-1-butanol in an electro-bioreactor. Carbon dioxide is the sole carbon source for this process and electricity is used as the energetic component. The process they developed effectively separates the light and dark reactions that occur during photosynthesis. Solar panels are used to convert sunlight to electrical energy which is then converted using the microorganism to a chemical intermediate. The team is now in the process of scaling up the operation and believes this process will be more efficient than the biologic process.

Improving Efficiency

In late 2012, a new discovery made the alternative fuel butanol more attractive to the biofuel industry. Scientist Hao Feng found a method that could significantly reduce the cost of the energy involved in making butanol. His team was able to isolate the butanol molecules during the fermentation process so they do not kill the organisms, and produces 100% or more butanol. After the fermentation process, they used a process called cloud point separation to recover the butanol which used 4 times less energy.

Also in late 2012, using systems metabolic engineering, a Korean research team at the former Korea Advanced Institute of Science and Technology (KAIST) has succeeded in demonstrating an optimized process to increase butanol production by generating an engineered bacterium. Professor Sang Yup Lee at the Department of Chemical and Biomolecular Engineering, KAIST, Dr. Do Young Seung at GS Caltex, a large oil refining company in Korea, and Dr. Yu-Sin Jang at BioFuelChem, a startup butanol company in Korea, applied a systems metabolic engineering approach to improve the production of butanol through enhancing the performance of Clostridium acetobutylicum, one of the best known butanol-producing bacteria. In addition, the downstream process was optimized and an in situ recovery process was integrated to achieve higher butanol titer, yield, and productivity. The combination of systems metabolic engineering and bioprocess optimization resulted in the development of a process capable of producing more than 585 g of butanol from 1.8 kg of glucose, which allows the production of this important industrial solvent and advanced biofuel to be cost competitive.

The anaerobic bacteria C. pasteurianum, C. acetobutylicum, and other Clostridium species have metabolic pathways that convert glycerol to butanol through fermentation. However, the production of butanol from glycerol by fermentation in C. Pasteurianum is low. To counter this, a group of researchers used chemical mutagenesis to create a hyper butanol-producing strain. The best mutant strain in this study "MBEL_GLY2" produced 10.8 g of butanol per 80 g of glycerol fed to the bacteria. This improvement compares to the 7.6 g butanol produced by the native bacteria.

Many organisms have the capacity to produce butanol utilizing an acetyl-CoA dependent pathway. The main problem with this pathway is the first reaction involving the condensation of two acetyl-CoA molecules to acetoacetyl-CoA. This reaction is thermodynamically unfavorable due to the positive Gibbs free energy associated with it (dG = 6.8 kcal/mol). Some experimentation has been done that involves increasing the carbon storage through the organism by utilizing carbon dioxide flow through photosynthetic organisms. To follow in this path of research, scientists have attempted to engineer reaction pathways that can enable photosynthetic organisms (like blue-green algae) to produce butanol more efficiently.

A study done by Ethan I. Lan and James C. Liao attempted to utilize the ATP produced during photosynthesis in blue-green algae to work around the thermodynamically unfavorable acetyl-CoA condensation to acetoacetyl-CoA. The native system was re-engineered to have acetyl-CoA react with ATP and CO_2 to form an intermediate, malonyl-CoA. Malonyl-CoA then reacts with another acetyl-CoA to form the desired acetoacetyl-CoA. The energy release from ATP hydrolysis (dG = -7.3 kcal/mol) makes this pathway significantly more favorable than standard condensation. Because blue-green algae generate NADPH during photosynthesis, it can be assumed that the cofactor environment is NADPH rich. Therefore, the native reaction pathway was further engineered to use NADPH rather than the standard NADH. All of these adjustments led to a 4-fold increase in butanol production, showing the importance of ATP and cofactor driving forces as a design principle in pathway engineering.

Producers

DuPont and BP plan to make biobutanol the first product of their joint effort to develop, produce, and market next-generation biofuels. In Europe the Swiss company Butalco is developing genetically modified yeasts for the production of biobutanol from cellulosic materials. Gourmet Butanol, a United States-based company, is developing a process that utilizes fungi to convert organic waste into biobutanol.

Distribution

Butanol better tolerates water contamination and is less corrosive than ethanol and more suitable for distribution through existing pipelines for gasoline. In blends with diesel or gasoline, butanol is less likely to separate from this fuel than ethanol if the fuel is contaminated with water. There is also a vapor pressure co-blend synergy with butanol and gasoline containing ethanol, which facilitates ethanol blending. This facilitates storage and distribution of blended fuels.

Table: Properties of Common Fuels:

Fuel	Energy density	Air-fuel ratio	Specific energy	Heat of vaporization	RON	MON	AKI
Gasoline and biogasoline	32 MJ/L	14.7	2.9 MJ/kg air	0.36 MJ/kg	91–99	81–89	87-95
Butanol fuel	29.2 MJ/L	11.1	3.6 MJ/kg air	0.43 MJ/kg	96	78	87
Anhydrous Ethanol fuel	19.6 MJ/L	9.0	3.0 MJ/kg air	0.92 MJ/kg	107	89	
Methanol fuel	16 MJ/L	6.4	3.1 MJ/kg air	1.2 MJ/kg	106	92	

Energy Content and Effects on Fuel Economy

Switching a gasoline engine over to butanol would in theory result in a fuel consumption penalty of about 10% but butanol's effect on mileage is yet to be determined by a scientific study. While the energy density for any mixture of gasoline and butanol can be calculated, tests with other alcohol fuels have demonstrated that the effect on fuel economy is not proportional to the change in energy density.

Octane Rating

The octane rating of n-butanol is similar to that of gasoline but lower than that of ethanol and methanol. n-Butanol has a RON (Research Octane number) of 96 and a MON (Motor octane number) of 78 (with a resulting "(R+M)/2 pump octane number" of 87, as used in North America) while t-butanol has octane ratings of 105 RON and 89 MON. t-Butanol is used as an additive in gasoline but cannot be used as a fuel in its pure form because its relatively high melting point of 25.5 °C (79 °F) causes it to gel and solidify near room temperature. On the other hand, isobutanol has a lower melting point than n-butanol and favorable RON of 113 and MON of 94, and is thus much better suited to high fraction gasoline blends, blends with n-butanol, or as a standalone fuel.

A fuel with a higher octane rating is less prone to knocking (extremely rapid and spontaneous combustion by compression) and the control system of any modern car engine can take advantage of this by adjusting the ignition timing. This will improve energy efficiency, leading to a better fuel economy than the comparisons of energy content different fuels indicate. By increasing the compression ratio, further gains in fuel economy, power and torque can be achieved. Conversely, a fuel with lower octane rating is more prone to knocking and will lower efficiency. Knocking can also cause engine damage. Engines designed to run on 87 octane will not have any additional power/fuel economy from being operated with higher octane fuel.

Air-fuel Ratio

Alcohol fuels, including butanol and ethanol, are partially oxidized and therefore need to run at richer mixtures than gasoline. Standard gasoline engines in cars can adjust the air-fuel ratio to accommodate variations in the fuel, but only within certain limits depending on model. If the limit is exceeded by running the engine on pure ethanol or a gasoline blend with a high percentage of ethanol, the engine will run lean, something which can critically damage components. Compared to ethanol, butanol can be mixed in higher ratios with gasoline for use in existing cars without the need for retrofit as the air-fuel ratio and energy content are closer to that of gasoline.

Specific Energy

Alcohol fuels have less energy per unit weight and unit volume than gasoline. To make it possible to compare the net energy released per cycle a measure called the fuels specific energy is sometimes used. It is defined as the energy released per air fuel ratio. The net energy released per cycle is higher for butanol than ethanol or methanol and about 10% higher than for gasoline.

Viscosity

Substance	Kinematic viscosity at 20 °C
Butanol	3.64 cSt
Diesel	>3 cSt
Ethanol	1.52 cSt
Water	1.0 cSt
Methanol	0.64 cSt
Gasoline	0.4–0.8 cSt

The viscosity of alcohols increase with longer carbon chains. For this reason, butanol is used as an alternative to shorter alcohols when a more viscous solvent is desired. The kinematic viscosity of butanol is several times higher than that of gasoline and about as viscous as high quality diesel fuel.

Heat of Vaporization

The fuel in an engine has to be vaporized before it will burn. Insufficient vaporization is a known problem with alcohol fuels during cold starts in cold weather. As the heat of vaporization of butanol is less than half of that of ethanol, an engine running on butanol should be easier to start in cold weather than one running on ethanol or methanol.

Potential Problems with the use of Butanol Fuel

The potential problems with the use of butanol are similar to those of ethanol:

- To match the combustion characteristics of gasoline, the utilization of butanol fuel as a substitute for gasoline requires fuel-flow increases (though butanol has only slightly less energy than gasoline, so the fuel-flow increase required is only minimal, maybe 10%, compared to 40% for ethanol).

- Alcohol-based fuels are not compatible with some fuel system components.

- Alcohol fuels may cause erroneous gas gauge readings in vehicles with capacitance fuel level gauging.

- While ethanol and methanol have lower energy densities than butanol, their higher octane number allows for greater compression ratio and efficiency.

- Butanol is one of many side products produced from current fermentation technologies; as a consequence, current fermentation technologies allow for very low yields of pure extracted butanol. When compared to ethanol, butanol is more fuel efficient as a fuel alternative, but ethanol can be produced at a much lower cost and with much greater yields.

- Butanol is toxic at a rate of 20g per liter and may need to undergo Tier 1 and Tier 2 health effects testing before being permitted as a primary fuel by the EPA.

Possible Butanol Fuel Mixtures

Standards for the blending of ethanol and methanol in gasoline exist in many countries, including the EU, the US and Brazil. Approximate equivalent butanol blends can be calculated from the relations between the stoichiometric fuel-air ratio of butanol, ethanol and gasoline. Common ethanol fuel mixtures for fuel sold as gasoline currently range from 5% to 10%. The share of butanol can be 60% greater than the equivalent ethanol share, which gives a range from 8% to 16%. "Equivalent" in this case refers only to the vehicle's ability to adjust to the fuel. Other properties such as energy density, viscosity and heat of vaporization will vary and may further limit the percentage of butanol that can be blended with gasoline. It was estimated that around 9.5 gigaliter (Gl) of gasoline can be saved and about 64.6 Gl of butanol-gasoline blend 16% (Bu16) can potentially be produced from corn residues in the US, which is equivalent to 11.8% of total domestic gasoline consumption.

Consumer acceptance may be limited due to the potentially offensive banana-like smell of n-butanol. Plans are underway to market a fuel that is 85% Ethanol and 15% Butanol (E85B), so existing E85 internal combustion engines can run on a 100% renewable fuel that could be made without using any fossil fuels. Because its longer hydrocarbon chain causes it to be fairly non-polar, it is more similar to gasoline than it is to ethanol. Butanol has been demonstrated to work in vehicles designed for use with gasoline without modification.

Pellet Fuel

Pellet fuels (or pellets) are biofuels made from compressed organic matter or biomass. Pellets can be made from any one of five general categories of biomass: industrial waste and co-products, food waste, agricultural residues, energy crops, and virgin lumber. Wood pellets are the most common type of pellet fuel and are generally made from compacted sawdust and related industrial wastes from the milling of lumber, manufacture of wood products and furniture, and construction. Other industrial waste sources include empty fruit bunches, palm kernel shells, coconut shells, and tree tops and branches discarded during logging operations. So-called "black pellets" are made of biomass, refined to resemble hard coal and were developed to be used in existing coal-fired power plants. Pellets are categorized by their heating value, moisture and ash content, and dimensions. They can be used as fuels for power generation, commercial or residential heating, and cooking. Pellets are extremely dense and can be produced with a low moisture content (below 10%) that allows them to be burned with a very high combustion efficiency.

Wood pellets.

Further, their regular geometry and small size allow automatic feeding with very fine calibration. They can be fed to a burner by auger feeding or by pneumatic conveying. Their high density also permits compact storage and transport over long distance. They can be conveniently blown from a tanker to a storage bunker or silo on a customer's premises.

A broad range of pellet stoves, central heating furnaces, and other heating appliances have been developed and marketed since the mid-1980s. In 1997 fully automatic wood pellet boilers with similar comfort level as oil and gas boilers became available in Austria. With the surge in the price of fossil fuels since 2005, the demand for pellet heating has increased in Europe and North America, and a sizable industry is emerging. According to the International Energy Agency Task 40, wood pellet production has more than doubled between 2006 and 2010 to over 14 million tons. In a 2012 report, the Biomass Energy Resource Center says that it expects wood pellet production in North America to double again in the next five years.

Production

Pellets are produced by compressing the wood material which has first passed through a hammer mill to provide a uniform dough-like mass. This mass is fed to a press, where it is squeezed through a die having holes of the size required (normally 6 mm diameter, sometimes 8 mm or larger). The high pressure of the press causes the temperature of the wood to increase greatly, and the lignin plasticizes slightly, forming a natural "glue" that holds the pellet together as it cools.

Pellets can be made from grass and other non-woody forms of biomass that do not contain lignin. A 2005 news story from Cornell University News suggested that grass pellet production was more advanced in Europe than North America. It suggested the benefits of grass as a feedstock included its short growing time (70 days), and ease of cultivation and processing. The story quoted Jerry Cherney, an agriculture professor at the school, stating that grasses produce 96% of the heat of wood and that "any mixture of grasses can be used, cut in mid- to late summer, left in the field to leach out minerals, then baled and pelleted. Drying of the hay is not required for pelleting, making the cost of processing less than with wood pelleting." In 2012, the Department of Agriculture of Nova Scotia announced as a demonstration project conversion of an oil-fired boiler to grass pellets at a research facility.

Pellet truck being filled at a plant.

Rice-husk fuel-pellets are made by compacting rice-husk obtained as by-product of rice-growing from the fields. It also has similar characteristics to the wood-pellets and more environment-friendly, as the raw material is a waste-product. The energy content is about 4-4.2 kcal/kg and moisture content is typically less than 10%. The size of pellets is generally kept to be about 6 mm diameter and 25 mm length in the form of a cylinder; though larger cylinder or briquette forms are not uncommon. It is much cheaper than similar energy-pellets and can be compacted/manufactured from the husk at the farm itself, using cheap machinery. They generally are more environment-friendly as compared to wood-pellets. In the regions of the world where wheat is the predominant food-crop, wheat husk can also be compacted to produce energy-pellets, with characteristics similar to rice-husk pellets.

A report by CORRIM (Consortium On Research on Renewable Industrial Material) for the Life-Cycle Inventory of Wood Pellet Manufacturing and Utilization estimates the energy required to dry, pelletize and transport pellets is less than 11% of the energy content of the pellets if using pre-dried industrial wood waste. If the pellets are made directly from forest material, it takes up to 18% of the energy to dry the wood and additional 8% for transportation and manufacturing energy. An environmental impact assessment of exported wood pellets by the Department of Chemical and Mineral Engineering, University of Bologna, Italy and the Clean Energy Research Centre, at the University of British Columbia, published in 2009, concluded that the energy consumed to ship Canadian wood pellets from Vancouver to Stockholm (15,500 km via the Panama Canal), is about 14% of the total energy content of the wood pellets.

Pellet Standards

Pellets conforming to the norms commonly used in Europe (DIN 51731 or Ö-Norm M-7135) have less than 10% water content, are uniform in density (higher than 1 ton per cubic meter, thus it sinks in water) (bulk density about 0.6-0.7 ton per cubic meter), have good structural strength, and low dust and ash content. Because the wood fibres are broken down by the hammer mill, there is virtually no difference in the finished pellets between different wood types. Pellets can be made from nearly any wood variety, provided the pellet press is equipped with good instrumentation, the differences in feed material can be compensated for in the press regulation. In Europe, the main production areas are located in south Scandinavia, Finland, Central Europe, Austria, and the Baltic countries.

Pellets conforming to the European standards norms which contain recycled wood or outside contaminants are considered Class B pellets. Recycled materials such as particle board, treated or painted wood, melamine resin-coated panels and the like are particularly unsuitable for use in pellets, since they may produce noxious emissions and uncontrolled variations in the burning characteristics of the pellets.

Standards used in the United States are different, developed by the Pellet Fuels Institute and, as in Europe, are not mandatory. Still, many manufacturers comply, as warranties of US-manufactured or imported combustion equipment may not cover damage by pellets non-conformant with regulations. Prices for US pellets surged during the fossil fuel price inflation of 2007–2008, but later dropped markedly and are generally lower on a per-BTU basis than most fossil fuels, excluding coal.

Regulatory agencies in Europe and North America are in the process of tightening the emissions standards for all forms of wood heat, including wood pellets and pellet stoves. These standards will become mandatory, with independently certified testing to ensure compliance. In the United States, the new rules initiated in 2009 have completed the EPA regulatory review process, with final new rules issued for comment on June 24, 2014. The American Lumber Standard Committee will be the independent certification agency for the new pellet standards.

Hazards

Wood pellets, in particular freshly made, are chemically active and can deplete the atmosphere of the oxygen required to sustain life. Wood pellets can also emit large quantities of the poisonous carbon monoxide. Fatal accidents have taken place in private storerooms and onboard marine vessels. When handled, wood pellets give off fine dust which can cause serious dust explosions.

Pellet Stove Operation

There are three general types of pellet heating appliances, free standing pellet stoves, pellet stove inserts and pellet boilers. *Pellet stoves* "look like traditional wood stoves but operate more like a modern furnace. Fuel, wood or other biomass pellets, is stored in a storage bin called a hopper. The hopper can be located on the top of the appliance, the side of it or remotely. A mechanical auger automatically feeds the pellets into a burn pot, where they are incinerated at such a high temperature that they create no vent-clogging creosote and very little ash or emissions "Heat-exchange tubes": Send air heated by fire into room. "Convection fan": Circulates air through heat-exchange tubes and into room. The biggest difference between a pellet stove and a woodstove, is that, inside, the pellet stove is a high-tech device with a circuit board, a thermostat, and fans—all of which work together to [regulate temperature and] heat your space efficiently."

A pellet stove insert is a stove that is inserted into an existing masonry or prefabricated wood fireplace.

Pellet boilers are standalone central heating and hot water systems designed to replace traditional fossil fuel systems in residential, commercial and institutional applications. Automatic or auto-pellet boilers include silos for bulk storage of pellets, a fuel delivery system that moves the fuel from the silo to the hopper, a logic controller to regulate temperature across multiple heating zones and an automated ash removal system for long-term automated operations.

Pellet baskets allow a person to heat their home using pellets in existing stoves or fireplaces.

Energy Output and Efficiency

Wood-pellet heater.

The energy content of wood pellets is approximately 4.7 – 5.2 MWh/tonne (~7450 BTU/lb), 14.4-20.3 MJ/kg.

High-efficiency wood pellet stoves and boilers have been developed in recent years, typically offering combustion efficiencies of over 85%. The newest generation of wood pellet boilers can work in condensing mode and therefore achieve 12% higher efficiency values. Wood pellet boilers have limited control over the rate and presence of combustion compared to liquid or gaseous-fired systems; however, for this reason they are better suited for hydronic heating systems due to the hydronic system's greater ability to store heat. Pellet burners capable of being retrofitted to oil-burning boilers are also available.

Air Pollution Emissions

Emissions such as NO_x, SO_x and volatile organic compounds from pellet burning equipment are in general very low in comparison to other forms of combustion heating. A recognized problem is the emission of fine particulate matter to the air, especially in urban areas that have a high concentration of pellet heating systems or coal or oil heating systems in close proximity. This $PM_{2.5}$ emissions of older pellet stoves and boilers can be problematic in close quarters, especially in comparison to natural gas (or renewable biogas), though on large installations electrostatic precipitators, cyclonic separators, or baghouse particle filters can control particulates when properly maintained and operated.

Global Warming

There is uncertainty to what degree making heat or electricity by burning wood pellets contributes to global climate change, as well as how the impact on climate compares to the impact of using competing sources of heat. Factors in the uncertainty include the wood source, carbon dioxide emissions from production and transport as well as from final combustion, and what time scale is appropriate for the consideration.

A report by the Manomet Center for Conservation Sciences, "Biomass Sustainability and Carbon Policy Study" issued in June 2010 for the Massachusetts Department of Energy Resources, concludes that burning biomass such as wood pellets or wood chips releases a large amount of CO_2 into the air, creating a "carbon debt" that is not retired for 20–25 years and after which there is a net benefit. In June 2011 the department was preparing to file its final regulation, expecting to significantly tighten controls on the use of biomass for energy, including wood pellets. Biomass energy proponents have disputed the Manomet report's conclusions, and scientists have pointed out oversights in the report, suggesting that climate impacts are worse than reported.

Until ca. 2008 it was commonly assumed, even in scientific papers, that biomass energy (including from wood pellets) is carbon neutral, largely because regrowth of vegetation was believed to recapture and store the carbon that is emitted to the air. Then, scientific papers studying the climate implications of biomass began to appear which refuted the simplistic assumption of its carbon neutrality. According to the Biomass Energy Resource Center, the assumption of carbon neutrality "has shifted to a recognition that the carbon implications of biomass depend on how the fuel is harvested, from what forest types, what kinds of forest management are applied, and how biomass is used over time and across the landscape."

In 2011 twelve prominent U.S. environmental organizations adopted policy setting a high bar for government incentives of biomass energy, including wood pellets. It states in part that, "biomass sources and facilities qualifying for (government) incentives must result in lower life-cycle, cumulative and net GHG and ocean acidifying emissions, within 20 years and also over the longer term, than the energy sources they replace or compete with."

Sustainability

The wood products industry is concerned that if large-scale use of wood energy is instituted, the supply of raw materials for construction and manufacturing will be significantly curtailed.

Cost

Due to the rapid increase in popularity since 2005, pellet availability and cost may be an issue. This is an important consideration when buying a pellet stove, furnace, pellet baskets or other devices known in the industry as Bradley Burners. However, current pellet production is increasing and there are plans to bring several new pellet mills online in the US in 2008–2009.

The cost of the pellets can be affected by the building cycle leading to fluctuations in the supply of sawdust and offcuts.

Per the New Hampshire Office of Energy and Planning release on Fuel Prices updated on 5 Oct 2015, the cost of #2 Fuel Oil delivered can be compared to the cost of Bulk Delivered Wood Fuel Pellets using their BTU equivalent: 1 ton pellets = 118.97 gallon of #2 Fuel Oil. This assumes that one ton of pellets produces 16,500,000 BTU and one gallon of #2 Fuel Oil produces 138,690 BTU. Thus if #2 Fuel Oil delivered costs $1.90/Gal, the breakeven price for pellets is $238.00/Ton delivered.

Usage by Region

Europe

EU Pellet Use (ton)	
Country	2013
UK	4 540 000
Italy	3 300 000
Denmark	2 500 000
Netherlands	2 000 000
Sweden	1 650 000
Germany	1 600 000
Belgium	1 320 000

Usage across Europe varies due to government regulations. In the Netherlands, Belgium, and the UK, pellets are used mainly in large-scale power plants. The UK's largest power plant, the Drax power station, converted some of its units to pellet burners starting in 2012; by 2015 Drax had made the UK the largest recipient of exports of wood pellets from the US. In Denmark and Sweden, pellets are used in large-scale power plants, medium-scale district heating systems, and small-scale residential heat. In Germany, Austria, Italy, and France, pellets are used mostly for small-scale residential and industrial heat.

The UK has initiated a grant scheme called the Renewable Heat Incentive (RHI) allowing non-domestic and domestic wood pellet boiler installations to receive payments over a period of between 7–20 years It is the first such scheme in the world and aims to increase the amount of renewable energy generated in the UK, in line with EU commitments. Scotland and Northern Ireland have separate but similar schemes. From Spring 2015, any biomass owners—whether domestic or commercial—must buy their fuels from BSL (Biomass Suppliers List) approved suppliers in order to receive RHI payments.

Pellets are widely used in Sweden, the main pellet producer in Europe, mainly as an alternative to oil-fired central heating. In Austria, the leading market for pellet central heating furnaces (relative to its population), it is estimated that $\frac{2}{3}$ of all new domestic heating furnaces are pellet burners. In Italy, a large market for automatically fed pellet stoves has developed. Italy's main usage for pellets is small-scale private residential and industrial boilers for heating.

In 2014 in Germany the overall wood pellet consumption per year comprised 2,2 mln tones. These pellets are consumed predominantly by residential small scale heating sector. The co-firing plants which use pellet sector for energy production are not widespread in the country. The largest amount of wood pellets is certified with DINplus and these are the pellets of the highest quality. As a rule, the pellets of lower quality are exported.

India

In 2019, India started co-firing biomass pellets in coal fired power stations around its capital city Delhi to reduce the air pollution caused by the stubble/biomass burning in open fields to clear the fields for sowing next crop. Plans are made to use biomass pellets for power generation through out the country to utilize nearly 145 million tonnes of agricultural residue to replace equal quantity of imported coal in power generation.

New Zealand

The total sales of wood pellets in New Zealand was 3–500,000 tonnes in 2013. Recent construction of new wood pellet plants has given a huge increase in production capacity.

United States

Some companies import European-made boilers. As of 2009, about 800,000 Americans were using wood pellets for heat. It is estimated that 2.33 million tons of wood pellets will be used for heat in the US in 2013. The US wood pellet export to Europe grew from 1.24 million ton in 2006 to 7 million ton in 2012, but forests grew even more.

Other Uses

Horse Bedding

When small amounts of water are added to wood pellets, they expand and revert to sawdust. This makes them suitable to use as a horse bedding. The ease of storage and transportation are additional benefits over traditional bedding. However, some species of wood, including walnut, can be toxic to horses and should never be used for bedding.

In Thailand, rice husk pellets are being produced for animal bedding. They have a high absorption rate which makes them ideal for the purpose.

Cattle Fodder

The biomass pellets made from edible matter can also be used as cattle fodder by importing from far away fodder surplus places to overcome the fodder shortage.

Absorbents

Wood pellets are also used to absorb contaminated water when drilling oil or gas wells.

Cooking

Wood pellet grills have gained popularity as a versatile way to grill, bake, and smoke. The size of the pellets makes it useful for creating a wood fired grill that still controls its temperature precisely.

Biodiesel

Biodiesel is an alternative fuel similar to conventional or 'fossil' diesel. Biodiesel can be produced from straight vegetable oil, animal oil/fats, tallow and waste cooking oil. The process used to convert these oils to Biodiesel is called transesterification. The largest possible source of suitable oil comes from oil crops such as rapeseed, palm or soybean. In the UK rapeseed represents the greatest potential for biodiesel production. Most biodiesel produced at present is produced from waste vegetable oil sourced from restaurants, chip shops, industrial food producers such as Birdseye etc. Though oil straight from the agricultural industry represents the greatest potential source it is not being produced commercially simply because the raw oil is too expensive. After the cost of converting it to biodiesel has been added on it is simply too expensive to compete with fossil diesel. Waste vegetable oil can often be sourced for free or sourced already treated for a small price. (The waste oil must be treated before conversion to biodiesel to remove impurities). The result is Biodiesel produced from waste vegetable oil can compete with fossil diesel.

Rapeseed fields produce vivid colours.

Benefits of Biodiesel

Biodiesel has many environmentally beneficial properties. The main benefit of biodiesel is that it can be described as 'carbon neutral'. This means that the fuel produces no net output of carbon in the form of carbon dioxide (CO_2). This effect occurs because when the oil crop grows it absorbs the same amount of CO_2 as is released when the fuel is combusted. In fact this is not completely accurate as CO_2 is released during the production of the fertilizer required to fertilize the fields in which the oil crops are grown. Fertilizer production is not the only source of pollution associated with the production of biodiesel, other sources include the esterification process, the solvent extraction of the oil, refining, drying and transporting. All these processes require an energy input either in the form of electricity or from a fuel, both of which will generally result in the release of green house gases. To properly assess the impact of all these sources requires use of a technique called life cycle analysis. Biodiesel is rapidly biodegradable and completely non-toxic, meaning spillages represent far less of a risk than fossil diesel spillages. Biodiesel has a higher flash point than fossil diesel and so is safer in the event of a crash.

Biodiesel Production

Biodiesel can be produced from straight vegetable oil, animal oil/fats, tallow and waste oils. There are three basic routes to biodiesel production from oils and fats:

- Base catalyzed transesterification of the oil.

- Direct acid catalyzed transesterification of the oil.

- Conversion of the oil to its fatty acids and then to biodiesel.

Almost all biodiesel is produced using base catalyzed transesterification as it is the most economical process requiring only low temperatures and pressures and producing a 98% conversion yield.

The Transesterification process is the reaction of a triglyceride (fat/oil) with an alcohol to form esters and glycerol. A triglyceride has a glycerine molecule as its base with three long chain fatty acids attached. The characteristics of the fat are determined by the nature of the fatty acids attached to the glycerine. The nature of the fatty acids can in turn affect the characteristics of the biodiesel. During the esterification process, the triglyceride is reacted with alcohol in the presence of a catalyst, usually a strong alkaline like sodium hydroxide. The alcohol reacts with the fatty acids to form the mono-alkyl ester, or biodiesel and crude glycerol. In most production methanol or ethanol is the alcohol used (methanol produces methyl esters, ethanol produces ethyl esters) and is base catalysed by either potassium or sodium hydroxide. Potassium hydroxide has been found to be more suitable for the ethyl ester biodiesel production, either base can be used for the methyl ester. A common product of the transesterification process is Rape Methyl Ester (RME) produced from raw rapeseed oil reacted with methanol.

The figure below shows the chemical process for methyl ester biodiesel. The reaction between the fat or oil and the alcohol is a reversible reaction and so the alcohol must be added in excess to drive the reaction towards the right and ensure complete conversion.

The products of the reaction are the biodiesel itself and glycerol.

A successful transesterification reaction is signified by the separation of the ester and glycerol layers after the reaction time. The heavier, co-product, glycerol settles out and may be sold as it is or it may be purified for use in other industries, e.g. the pharmaceutical, cosmetics etc.

Straight vegetable oil (SVO) can be used directly as a fossil diesel substitute however using this fuel can lead to some fairly serious engine problems. Due to its relatively high viscosity SVO leads to poor atomisation of the fuel, incomplete combustion, coking of the fuel injectors, ring carbonisation, and accumulation of fuel in the lubricating oil. The best method for solving these problems is the transesterification of the oil.

The engine combustion benefits of the transesterification of the oil are:

- Lowered viscosity

- Complete removal of the glycerides

- Lowered boiling point

- Lowered flash point

- Lowered pour point

Production Process

An example of a simple production flow chart is proved below with a brief explanation of each step.

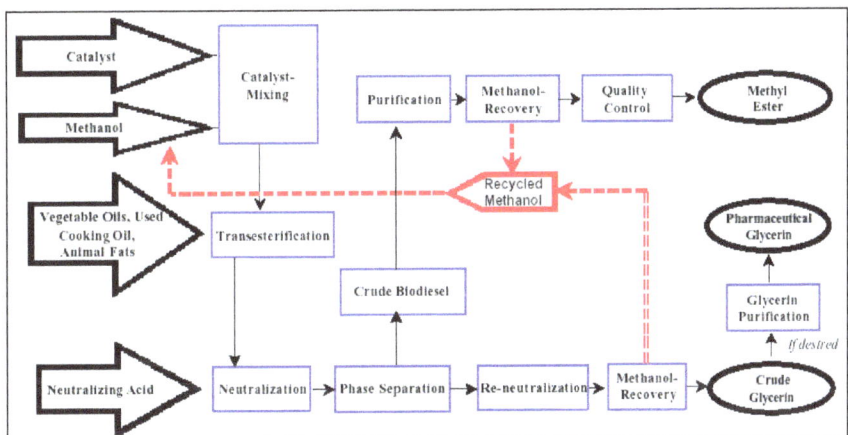

Mixing of Alcohol and Catalyst

The catalyst is typically sodium hydroxide (caustic soda) or potassium hydroxide (potash). It is dissolved in the alcohol using a standard agitator or mixer reaction. The alcohol/catalyst mix is then charged into a closed reaction vessel and the oil or fat is added. The system from here on is totally closed to the atmosphere to prevent the loss of alcohol. The reaction mix is kept just above the boiling point of the alcohol (around 160 °F) to speed up the reaction and the reaction takes place. Recommended reaction time varies from 1 to 8 hours, and some systems recommend the reaction take place at room temperature. Excess alcohol is normally used to ensure total conversion of the fat or oil to its esters. Care must be taken to monitor the amount of water and free fatty acids in the incoming oil or fat. If the free fatty acid level or water level is too high it may cause problems with soap formation and the separation of the glycerin by-product downstream.

Separation

Once the reaction is complete, two major products exist: glycerin and biodiesel. Each has a substantial amount of the excess methanol that was used in the reaction. The reacted mixture is sometimes neutralized at this step if needed. The glycerin phase is much more dense than biodiesel phase and the two can be gravity separated with glycerin simply drawn off the bottom of the settling vessel. In some cases, a centrifuge is used to separate the two materials faster.

Alcohol Removal

Once the glycerin and biodiesel phases have been separated, the excess alcohol in each phase is removed with a flash evaporation process or by distillation. In others systems, the alcohol is removed and the mixture neutralized before the glycerin and esters have been separated. In either case, the alcohol is recovered using distillation equipment and is re-used. Care must be taken to ensure no water accumulates in the recovered alcohol stream.

Glycerin Neutralization

The glycerin by-product contains unused catalyst and soaps that are neutralized with an acid and sent to storage as crude glycerin. In some cases the salt formed during this phase is recovered for use as fertilizer. In most cases the salt is left in the glycerin. Water and alcohol are removed to produce 80-88% pure glycerin that is ready to be sold as crude glycerin. In more sophisticated operations, the glycerin is distilled to 99% or higher purity and sold into the cosmetic and pharmaceutical markets.

Methyl Ester Wash

Once separated from the glycerin, the biodiesel is sometimes purified by washing gently with warm water to remove residual catalyst or soaps, dried, and sent to storage. In some processes this step is unnecessary. This is normally the end of the production process resulting in a clear amber-yellow liquid with a viscosity similar to petrodiesel. In some systems the biodiesel is distilled in an additional step to remove small amounts of color bodies to produce a colorless biodiesel.

Product Quality

Prior to use as a commercial fuel, the finished biodiesel must be analyzed using sophisticated analytical equipment to ensure it meets any required specifications. The most important aspects of biodiesel production to ensure trouble free operation in diesel engines are:

- Complete Reaction

- Removal of Glycerin

- Removal of Catalyst

- Removal of Alcohol

- Absence of Free Fatty Acids

References

- Burton, George; Holman, John; Lazonby, John (2000). Salters Advanced Chemistry: Chemical Storylines (2nd ed.). Heinemann. ISBN 0-435-63119-5

- "First Commercial Plant". Carbon Recycling International. Archived from the original on 3 July 2013. Retrieved 11 July 2012

- Generations-of-biofuels-v1.3, bioenergy: oregonstate.edu, Retrieved 23 June, 2019

- "Technology". Carbon Recycling International. 2011. Archived from the original on 17 June 2013. Retrieved 11 July 2012

- Ethanol-fuel: conserve-energy-future.com, Retrieved 15 May, 2019

- "About the Densified Biomass Fuel Report". U.S. EIA. October 17, 2018. Retrieved October 23, 2018

- What-biodiesel, biofuels: esru.strath.ac.uk, Retrieved 16 June, 2019

PERMISSIONS

We would like to thank the editorial team for lending their expertise to make the book truly unique. They have played a crucial role in the development of this book. Without their invaluable contributions this book wouldn't have been possible. They have made vital efforts to compile up to date information on the varied aspects of this subject to make this book a valuable addition to the collection of many professionals and students.

This book was conceptualized with the vision of imparting up-to-date and integrated information in this field. To ensure the same, a matchless editorial board was set up. Every individual on the board went through rigorous rounds of assessment to prove their worth. After which they invested a large part of their time researching and compiling the most relevant data for our readers.

The editorial board has been involved in producing this book since its inception. They have spent rigorous hours researching and exploring the diverse topics which have resulted in the successful publishing of this book. They have passed on their knowledge of decades through this book. To expedite this challenging task, the publisher supported the team at every step. A small team of assistant editors was also appointed to further simplify the editing procedure and attain best results for the readers.

Apart from the editorial board, the designing team has also invested a significant amount of their time in understanding the subject and creating the most relevant covers. They scrutinized every image to scout for the most suitable representation of the subject and create an appropriate cover for the book.

The publishing team has been an ardent support to the editorial, designing and production team. Their endless efforts to recruit the best for this project, has resulted in the accomplishment of this book. They are a veteran in the field of academics and their pool of knowledge is as vast as their experience in printing. Their expertise and guidance has proved useful at every step. Their uncompromising quality standards have made this book an exceptional effort. Their encouragement from time to time has been an inspiration for everyone.

The publisher and the editorial board hope that this book will prove to be a valuable piece of knowledge for students, practitioners and scholars across the globe.

INDEX

www.ingramcontent.com/pod-product-compliance
Lightning Source LLC
Chambersburg PA
CBHW080403190526
45161CB00003B/120